The language of environment
A *new rhetoric*

George Myerson
King's College London

Yvonne Rydin
London School of Economics and Political Science

UCL
PRESS

First published in 1996 by UCL Press.

UCL Press Limited
University College London
Gower Street
London WC1E 6BT

The name of University College London (UCL) is a registered
trade mark used by UCL Press with the consent of the owner.

ISBNs:1-85728-330-9 HB
1-85728-331-7 PB

British Library Cataloguing in Publication Data
A catalogue record for this book is available from the British Library.

Typeset in Classical Garamond.
Printed and bound by
Bookcraft (Bath) Ltd, England.

Contents

Contents

Preface

"Environment" challenges modern knowledge and its institutions: academic disciplines, research groups, journals and presses, syllabuses and texts, professions and data banks, media experts and policy advisors. The challenge is twofold. First, "environment" does not fit the divisions of modern specialization. In academic terms, "environment" belongs to every discipline and to none: chemistry and biology, ecology and sociology, philosophy and geography, engineering and politics, psychology and history, media studies and cultural theory. In media terms, it is harder and harder to draw a line around certain issues and say: this is an environmental issue. War and trade, food and transport, weather and protest, animal rights and international relations: "environment" is inescapable, and it challenges the boundaries between news areas that are otherwise distinct, domestic or international, major catastrophe or minor incident, good news or bad news. "Environment" is beyond the disciplines of modern knowledge, and beyond the grid of modern media expertise. This book is an attempt to understand the cultural presence of environment, its peculiar and particular "interfusion", to adapt Wordsworth's "sense sublime/Of something far more deeply interfused". How has environment come to pose its profound challenge, a challenge not to one area of knowledge but to our whole conception of knowledge?

This takes us to the second challenge of environment: how can we use all our knowledge and ideas, in the face of a possible or potential environmental crisis? Which knowledge will help? And how do we begin to rethink the role of diverse fields? How does a culture that has invested so much in ever more complex specialization now address an agenda that transgresses all the divisions, science and arts, objectivity and feeling, experiment and ethics, forecasting and mythology? *The language of environment* makes no policy proposals, it is not prescriptive. But it is an attempt to think about the cultural context of all proposals and prescriptions, the cultures of authority and expertise in our time. How is knowledge made to count, and how do all the different claims connect, or collide?

In one way, there is something provocative about "language of environment": isn't environment above or beyond language? Must we linger over the words, when the subject is so important? But all the knowledge and all the proposals, all the expertise and experience, the values and visions, they only act on society in terms of language. If there are no words, there are no impacts.

v

Preface

And it is because environment has impacted so profoundly on contemporary language, that the challenge is so pervasive and deeply felt across such diverse areas. We like to think of specialized disciplines, each following their own necessary course, responding to discoveries and research developments, scholarly debates and theories. The story is partly true. But there is another perspective too, the cultural perspective in which specialisms are responses to a more diffuse and general agenda of questions. By "language of environment", we also mean the clusters of image and story by which that general culture impacts on all the specialized fields. Specialized knowledge needs to use current language in order to have an influence; it is also a response to current language, in the broadest sense.

For language is more than words, and the language of environment is more than environment words. We do concentrate around certain key "environment words", but in order to hear the voices which write and speak with them, the arguments that are made through them. Language is our consciousness of the world. If environment represents a changed consciousness of the world, then environment must also be a changed language. We want to explore how deep the change of cultural consciousness has gone.

The language of environment is not a review of environmental debates; it is an active engagement with them, not a polemical engagment but an involved inquiry. This inquiry necessarily overlaps discipines, as its subjet matter does. It is, in some sense of the word, an "interdisciplinary" text. This of course raises the issue of what is meant by "interdisciplinary".

There is one, fairly conventional rendering of the term which suggests two or more disciplines coming together, bringing their own skills, knowledge and other professional baggage to a conversation which takes place on borders or other neutral ground. Each then retreats to their own disciplinary space, reinforced with a bit more new knowledge and insights from the frontier of that space. This book does not fit with this model.

There is another model, also conventional in its own arena, which sees a grander future for interdisciplinary work. Here interdisciplinarity becomes a field of work in its own right. It occupies the space between disciplines, the large background of deep space against which the discplinary solar systems exist. And it develops its own theories and knowledge which exists across and above disciplinary theories and knowledge. Interdisciplinarity produces super-theories. Structuralism and marxism come to minds as examples of this model. Again, this book does not fit with this model.

Rather we hope to have produced a text which is part of a growing development, to which cultural geography say testifies, that the disciplines themselves are not homogeneous. There is already dialogue occuring within each discipline, dialogues between different kinds of insight and between different claims to knowledge. Therefore there is no radical difference between disciplinary and interdisciplinary work. Both involve dialogues between intellectual stances and it is more fruitful to think of academic work as engagement with a variety of these dialogues. We see this in the "field" of cultural geography which is simultaneously part of geography, cultural studies, politics and a separate entity on its own. Again, environmental studies is a set of its own dialogues as well as belonging to geography, ecology, economics and sociology.

So our book is itself a dialogue with these existing sets of dialogues. This is not to make a weak claim. We argue that a dialogue of engagement is itself a source of knowledge and validity claims. After a seminar given at an early stage of development of the book, an Oxford academic said "Oh, so the point is to read texts more intelligently!" That is indeed our hope. Reading is essentially a personal activity; it cannot be done for someone else. Neither do we offer a mechanism or recipe for doing so. We seek to provide a path by example to engaging with texts and the dialogues within them.

In doing so it is our belief that encouraging the richest set of dialogues is the best way to keep environmental issues on the political and social agenda. And it is also a way of forcing, prompting and shaping change. Stating one's own view forcefully is not necessarily the most effective way to achieve change. But in a democracy, understanding and making dialogues is itself a powerful resource.

Acknowledgements

There are a number of people we would like to thank for help in the complex technical process of assembling such a text with an unusual wealth of quoted material: Graham Calvert, Jonathan Krueger, Mark Pennington and Carole Webb helped with collation of the Profile and CD-rom material; Jonathan also undertook the unenvieable task of checking all the quotes; Roger Jones and his staff at UCL Press processed the text with great efficiency and creative lateral thinking when it came to annotating the quotes; David Ryder of the LSE Drawing Office produced the illustrations; and Christine Sellgren typed the bibliography with unerring accuracy. For joining in the creative process of writing and arguing about the writing, we wish to thank Jacqueline Burgess and Jim Throgmorton.

There are many wider intellectual debts and engagements. We are grateful to colleagues from LSE, UCL and Turin with whom we collaborated on a CEC study of *The economic and cultural conditions of the sustainable city*. It is hard to acknowledge conversations, as if they were texts, but since we have interpreted texts as dialogues, we ought to try: reading groups formed by Sonia Livingstone and Peter Lunt, and by Jacqueline Burgess's "Society and Nature" group; discussions arising around the Penn State Rhetoric and Composition Conference of 1994, creatively hosted by Don Bialostosky of Penn State, with many people, and particularly with Susan Miller of the University of Utah; engagement with the ideas and planning theory of Patsy Healey; rhetorical dialogues with Paul Kenny and economic utopias and realities with Simon Zadek. The whole project is surrounded by dialogue with Simon and Eleanor. Other voices run deep, "unheard melodies" still.

We acknowledge permission kindly given for the use of copyright material: Figure 5.1, Mrs Pestel; Figure 5.2, after diagram by Herman Daly, in *Steady state economics*, Earthscan, London 1992; Figure 5.3, after Lawrence Zeegen, reproduced from *Wealth beyond measure*, by Paul Ekins, Gaia Books Ltd ©, London 1992; Figure 5.4, after diagram by David Pearce; Figure 5.5, Hazel Henderson; extract from Ken Edwards *Good science*; Roof Books; extract from Ted Hughes *Creation*, Faber & Faber Ltd.; extract from Olive Senior *Gardening in the tropics*, McCelland & Stewart Inc.

CHAPTER ONE

Environment words, environet
and the rhetorical web

Environment words and environet

Anything *can* be argued about, but only some things *are* argued about, a fact easily taken for granted. Almost automatically, naturally, we discuss certain issues, the issues of the day and of the times. People recognize what is topical; this is "cultural literacy". We recognize the agenda, the viewpoints and key words. We speak for ourselves, but we also know what is generally being said and how to say it. Cultural literacy is consciousness of what is topical and how to be topical. It is reflected in talk, but it also applies to writing and reading, as much as to talking. Such literacy is widespread within society and is highly developed by specialists. Anyone trying to write an article for a newspaper, a magazine or an academic journal has an acute sense of what we are meant to be discussing now, even if the aim is to change the subject. We know the topics, and the turns of phrase, as talkers, as readers, as writers. But there is knowing and *knowing*: being aware of what one knows "naturally" is reflexive insight (Giddens 1994). This book encourages reflexive insight into environmental discussion, and specifically insight into the language people use to write about the environment, and how that language transmits feelings, shapes ideas, and connects visions.

It is difficult to be reflexive about what we do naturally and what we appear to know without effort. Ironically, such knowledge is hard, because its object is why we do certain things so easily. Reflexivity is a layer of thinking that contemporary life increasingly requires (Giddens 1992, 1994). But why should we think about what we do without thinking about it? And, in particular, why should we think reflexively about the environment, the environment seems the most natural of all topics: it is "out there", a physical entity, a real problem with real solutions – or not – real facts – or not. It is, surely, not like, say, economic debate, where the whole subject seems verbal, at some level, and where the issues are not separable from the words we use about them. Does "marginal utility" – the concept of an individual gaining more satisfaction from one more unit of consumption – exist prior to the economist's definition of it? This is even more of an issue with some moral questions: how could one discuss "freedom" or "rights" without having a language to frame the question in the first place? But "global warming" is not like either "inflation" or "the right to silence": global warming is a physical phenomenon: the temperature rises, or does not rise, the causes are either understood or not understood. We read and write about the environment, because of its material reality, because there is something evidently there to address. The agenda is doubly natural: we recognize it culturally, and we feel it around us.

At this point, authors sometimes announce a twist: "but we shall show that the agenda is far from natural, it is foisted upon us/you, beware you are being manipulated. Distrust the knowledge of what to discuss and how." We do not announce a twist, an unmasking. We do not see "reflexive insight" as an *antidote* to ordinary thought; on the contrary, without ordinary understanding there is nothing to reflect with or on (Aristotle 1926, Billig 1995). To think about how environmental texts are written does not free us from some wicked spell. We aren't uncovering hidden traps, showing how audiences are mis-led, tricked by clever and false words: "rhetoric" in one degraded sense. We are exploring a world, a world that comes into being through discussion, the discussion of real things, real problems, real crises, losses and remedies.

Our theme is *meaning*, how culture gives meanings to the worlds we inhabit. Environmental discussion has many functions: to protest, to expose, to reassure, to propitiate. And, as a result, laws are enacted, rules are revised, institutions are created and destroyed, lives endangered and saved. At the same time, meanings are created, thickened, discarded. And the meanings rebound, they affect the outcomes, the laws, rules and institutions. Indeed, the meanings become

Environment words and environet

the "situation", the cultural moment in which the environment is discussed – sympathetic to environmentalism, or wearied of it, anxious about pollution or inured to it, hopeful of improvement or cynical. Environmental arguments are factual, informative, often scientific. But they are also meaningful, suggestive and atmospheric.

Consider an encounter between opposing viewpoints, a staged debate took place in 1992 at the Columbia University of between two influential voices in the environmental debate: Norman Myers (a campaigning intellectual) and Julian Simon (an academic economist), reported in *Scarcity or abundance?* (1994). Julian Simon defends growth, industrial production and the market system; Norman Myers sees danger, a world at risk in contemporary trends. Simon is reassuring: "The picture also is now clear that population growth does not hinder economic development." (ibid.: 114–15). The phrasing matters: "does not hinder" leaves room for interpretation. In reply, Myers evokes a crisis in precise terms:

> And at the same time our Earth has taken on board another 93 million people, equivalent to more than a "new Mexico" – and this at a time when our Earth is straining under the burden of its present population of 5.5 billion people. (ibid.: 126–7)

Facts abound and, at the same time, meanings flow and collide. Simon's world is expandable; Myers's world is bounded, an ark at sea. Simon advances ". . . the theory which I believe explains how these good things can all be happening at once." (ibid.: 134). Myers replies: "To reiterate, I believe there is much evidence we are at a breakpoint in the human enterprise." (ibid.: 142).

Claim counters claim, but, ultimately, one language encounters another, and one self-presentation challenges another. Simon synthesizes: economic growth and population growth goes together; good things naturally combine. Myers disrupts the harmony by threatening another Mexico every year. (And why *Mexico*? What are the relevant associations here?) Simon's self-presentation is reassuring and energetic; Myers's self-presentation is urgent and alarmed, yet also objective and definitive. The contrast involves whole human personalities. The facts exist in relation to those personae, and we cannot choose between the facts without being involved in the personalities, and their languages. Whom do you admire, which would

3

you trust? The "whom" is not personal; the audience must choose between different worlds, different ways of life. The choice is about meanings and cultures.

Myers & Simon each offer a distinctive, personal voice, but they do not start from scratch. They draw on available facts, theories and examples. But it is more than that: phrases are ready-to-use forms of argument, even styles of personality. Since the late 1980s environmental issues have received a tremendous influx of controversial energy; it is this energy that we *feel* as we argue or read others' arguments, the push of possible things to say. Discourse – the totality of things written and read, spoken and heard – has flowed towards the environment, and it has not gone away. The environment remains topical: in this book we explain how that "topicality" works, at the level of language. Of course, from one point of view, topicality is attributable to events, things happening: the Chernobyl explosion, the "discovery" of the ozone hole, evidence on global warming, oil spills at sea. Institutions react, and people actively make new political agendas: the Brundtland Commission's work, leading on to the United Nations Conference on Environment and Development at Rio in 1992, Agenda 21 and the UN Commission on Sustainable Development.

Things are topical. But topicality has an interior dimension too, a cultural dimension. Language itself renews discussion and moves the agenda. It feels natural to write and read about a subject; it feels possible to talk about it in a certain way; there are ready ways to put a case. Topicality also means new ways to write, new idioms, new words. Environment is discussed because things happen, have happened, may happen; it is also discussed because arguments are available, rich arguments, and new cases keep arriving. Attention is waiting to be grabbed, and we already know how to grab it: the words are there. Books, articles, policy documents, pamphlets – all sorts of texts abound, specialist and public, official and personal, philosophical and strategic. Where do they all come from? How are they all writable and readable? Why do they not run dry, but on the contrary boil up again and again? How are *new* texts on the environment always possible today? What supports the production of so much new arguing? The questions are cultural, and require cultural analysis, a new analysis.

Furthermore, the world is at issue in environmental texts and their arguments. Three extracts illustrate this:

4

If we do not stem the proliferation of the world's deadliest weapons, no democracy can feel secure. If we do not strengthen the capacity to resolve conflicts among and in nations, those conflicts will smother the birth of free institutions, threaten the development of entire regions and continue to take innocent lives. If we do not nurture our people and our planet through sustainable development, we will deepen conflict and waste the very wonders that make our efforts worth doing. (President Clinton, reported 27 September 1993 by Associated Press)

The ecological teaching of the bible is simply inescapable: God made the world because He wanted it made. He thinks the world is good, and He loves it. It is His world; He has never relinquished title to it. And He has never revoked the conditions, bearing on His gift to us of the use of it, that oblige us to take excellent care of it. (Berry 1990: 98)

Since the Montreal Protocol was signed in 1987, scientists have found that reductions on CFC use called for were not nearly enough to save the ozone layer. They are seeking a total worldwide phase out as quickly as possible. (*Independent*, 15 June 1990)

In these quotations, the planet needs nurturing; the world wants care; science calls for a "worldwide" strategy: politics, religion and chemistry are invoked. These examples of oration, essay and expert evidence are all voices on what we shall term the environet. But why does "the world" present itself so compellingly in these texts, why does the environment make such strong texts nowadays? In this book, we address the questions in terms of language and how language drives arguments forwards. When one says "language drives", that does not mean people are passive, at the mercy of linguistic structures. We are studying invention, cultural invention and creativity as present in our use of language. People change language, and language changes us. Language can mean many things: abstract structures, rules, forms; in this book, we start from *the words*, not as in a dictionary, but words in voices, and voiced words, voices in relation to other voices. Myers & Simon are cultural voices, so is President Clinton (indeed he is many voices), and so are Wendell Berry and the nameless scientists.

We look specifically at voices from texts. By texts we mean any instance of talk, argument, discourse in a written form and we draw on poetry, newspaper articles, magazines, books and policy documents. However the term can and is used more broadly to indicate any object which can be "read", interpreted for its meanings and the interrelations between different viewpoints, lines of argument and use of imagery. Thus a text can be a radio or television programme (Livingstone & Lunt 1993), news or fiction (Leith & Myerson 1989), a film (Harvey 1989), a building or even a city (Dear 1995). And the text refers to both the part and the whole; in our case, where we quote many extracts, out texts are both the specific extracts and the larger document they are drawn from. A close metaphor is the religious sermon whose text is both the portion of the Bible under scrutiny and the argument of the whole speech by the preacher. Across the texts run what we call "environment words". To our approach, these words are central. The environment words are markers in the texts. Their presence reveals the environmental agenda. As they apprear, change and are contested, they show the dynamics of that agenda. More important, the words carry the viewpoints; without these marker words there would be no focus for the viewpoints. Imagine the debate about the rainforests without the environment word "biodiversity". Different voices, expressing these viewpoints, use the words, are heard through the words. By looking at enviroment words we come to hear the different voices of the environmental agenda. And as words engage with each other in argument, so the voices engage. The multiplicity of environment words, their repeatedly changing nature, the continuing quest to define their meanings are all central to the feel of contemporary environmental debates and to the feeling of present times. The arguments multiply, the words change, meanings are contested, not settled.

The words come alive in the voices, voices that contend, "criticize or uphold an argument" (Aristotle 1926: 3). The voices speak from texts, each text containing one or more voices, depending on how it is organized. For instance, Simon has an "expert" voice, an economist voice, and he also has a "forceful person" voice; Myers has a "biological" voice and an "anxious citizen" voice. The environment words leave their traces in text after text. As we have said, "text" means many things: it may be specialist, a monograph on biodiversity; or it may be a newspaper article on summer heatwaves and global warming; it may be polemical, denouncing environmentalists or developmentalists; it may be personal, a journey across South American rainforests; or it may be a catalogue of facts collected from different sources, an information collage about the climate or resource depletion. The text may be authorized by the UN or peer reviewed in a discipline; it may be vetted by a publisher or constructed by a political party. The environment words cross official policy documents and expert reports, news items, poems and statistics, presidential speeches and bestsellers, political theories and newspaper headlines.

The texts vary, shorter and longer, denser and simpler, more direct and more oblique. But each has a topical consciousness of the environment, of what currently should be argued about and how. Our metaphor for the aggregate collection of texts, words and voices is "the environet", a network making linkage upon linkage between the environment words. Such a network is a system for correcting elements with the language of environment, identifying stronger and weaker connections. However, it is only a metaphor for how language works; we do not suggest an object for network analysis (Dowding 1995).

The environet can also be imagined as a "textual carnival" in the metaphor of Susan Miller (1991). A carnival is also dynamic, full of connections made and broken in the melée. And a carnival (such as the Rio Earth Summit?) mixes people and voices: the prestigious and peripheral, authoritative and popular, would-be weighty and knowingly transient. So, the environet is also a dynamic system of changing connections spread across society. And this net is busy with environmental arguing, competitive and collaborative, controlled and spontaneous. "Environet" is a textual carnival for contemporary times, where high and low can exchange views and change places.

Each text comes from the environet, and adds to it. However, it is not necessarily *easy* to distinguish the voices (Box 1.1): is "a good atmosphere" so radically different from "the great smog switch-off"? The media articles convey specialist information, a university report and a NASA project; the academic article uses everyday phrases; a poem can mix science and politics in a phrase.

It was once asserted that modern culture would be dominated by "the image" and that the word would decline (McCluhan 1967). Certainly, images are everywhere and a different study could analyze the environet in terms of its images. Another study could look at

7

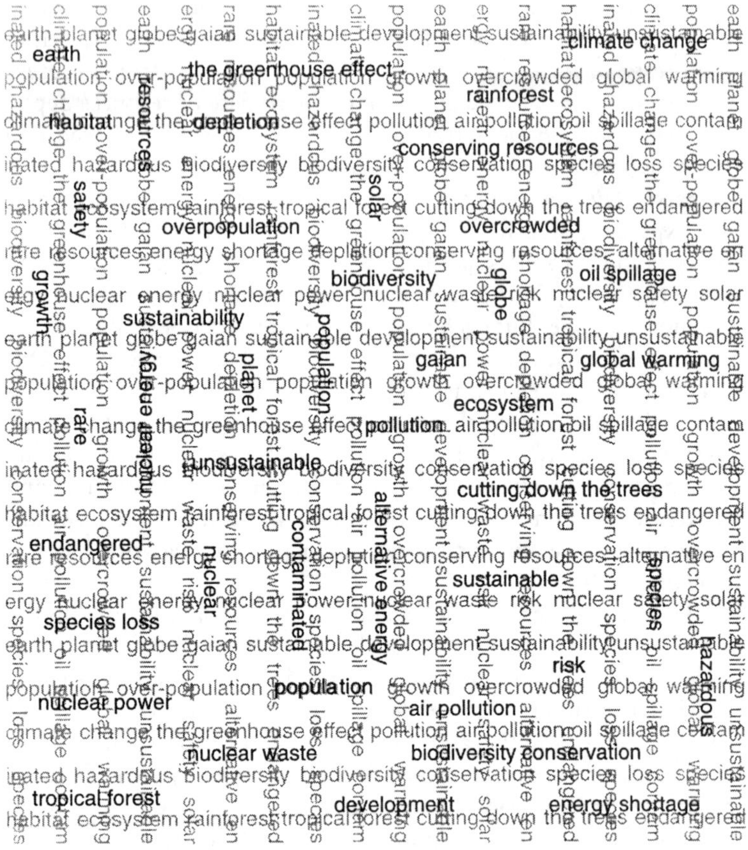

Figure 1.1 Environment words and the environet.

speech, everyday conversation and the environet of words and issues. But alongside image and talk, a sea of texts exists. In modern culture, communication is multiplied and diverse. Any argument cuts across many different media. Culture is made by texts, although not only by texts. Texts are one map of culture, and one expression of its contemporary riddle: what is everywhere linked and always fragmenting (Bauman 1992, Connor 1992). For behind this notion of environet is the analysis of society as structured by cultural practices, but only temporarily. The net holds in place the practices of communication

---

BOX 1.1   Voices on the environet

"Within
The incipient rain-forest in an under-construction bank foyer
The juicy realistic look of nature takes shape; the phone network
Evolves an ecology of its own, but crashing all the time . . .
The lakes have disappeared;
Coded with natural colours of the visible spectrum,
A local ecology of gardens now adorns
The homes of the privileged, salvaged from
The destroyed series. Lawns and golf courses . . ."
(Edwards, *A new word order*, 1992: 63, 70)

"The report, by Lancaster University's Centre for the Study of Environmental Change and the consultants Environmental Resources Management, says there is considerable local resentment at the perceived withholding of information about leaks and other pollution incidents at Sellafield." (*Financial Times*, 16 October 1993)

"Environment – The Great Smog Switch Off – Germans Put Air Quality to the Test" (*Guardian*, 13 August 1993)

"The National Aeronautics and Space Administration, a target for 'reinvention' ever since the Clinton administration arrived, is mentioned in the review as a place to improve contracting practices, strengthen management and clarify the objectives of 'Mission to Planet Earth', which will use a fleet of satellites to study environmental change." (*Washington Post*, 8 September 1993)

"Creating a Good Atmosphere: Minimum Participation for Tackling the 'Greenhouse Effect'" (Black et al. 1993)

---

and arguing, but is itself continually pushed into new shapes – stretched and torn in places, made slack and compressed in others. It is a representation of a view of modern culture as differentiated and connected by the same processes (Habermas 1984, 1987; Giddens 1990, 1992, 1994).

Sometimes, one text within the environet answers another, directly: a writer denounces a rival, criticizes and quotes, interprets and praises. Many texts accumulate around, say, the Brundtland Report (WCED 1987). But direct comment is the exception, not the rule. In the environet, texts always interact with each other, indirectly because they address issues together, they use the same words, they address similar audiences, they respond to similar feelings, they use the same arguments, or opposing arguments (Bakhtin 1986, Leith & Myerson 1989). The effect of these repeated connections is cumulative, building up support for different viewpoints and pushing certain issues into more prominent positions in the environmental agenda. As arguments accumulate across the environet, the agendas reform. Therefore, any single text has broader impacts than are felt by its readers; each contributes to the aggregate impacts of environmental texts on the culture as a whole. Someone reads an editorial on marine pollution; then an article on resource management for a formal seminar; then a review of a new book on Gaian philosophy. Even though the texts appear distinct and make no overt allusion to each other, there are cultural connections, and in the reading mind they fit together. Texts are collaged, stuck together to make picture after picture (Derrida 1992).

Of course, texts differ. Arguments derive from one context or another, be it instant news or learned journal. The word "text" does not mean "just texts", all the same with no distinctions. Texts are better and worse, more knowledgeable and less, more gripping and less, more complex and less.

Most people do not trust the government to look after the environment. More than 80 per cent believe the Tories do not care enough for nature. Ninety per cent fear for the planet with deforestation and traffic pollution the greatest worries, a Friends of the Earth survey revealed. (*Today*, 30 December 1993)

The use of energy accounts for a major fraction of all anthropogenic emissions of greenhouse gases (Intergovernmental Panel on Climate Change 1990), and in most industrialized countries the use of transportation and electricity accounts for a major fraction of all energy-related emissions. (DeLuchi 1993: 187)

Both pieces are topical. At the same time, each uses the topicality differently. The arguments may try to ignore each other, but they are interacting: "traffic pollution . . . worries", "energy related emissions".

The environment is arguable in a particular sense. Here "arguable" does not mean "doubtful"; it means capable of being discussed, rich in topicality. We shall explore where environmental argument leads, and what outcomes are possible. We are not seeking to grade the arguments, to decide right and wrong; we interpret the arguments, their structure and their interconnection. We reveal the *processes* of arguing (Habermas 1984), the words shifting, the ideas being connected, and the linking of facts with values, of information with identity. Our claim is that the argumentative environet is a central fact of contemporary culture.

Because environmental issues excite argument from different contexts, environmental arguing could transform culture, not alone but in a net of nets. For the environet is one among many nets of topicality, such as a "rightsnet". Like the others, the environet is also unique. It links sciences, social sciences, political manifestos, local news, and UN strategies. Ultimately, we shall suggest, the environet might transform contemporary culture. It could recreate argument itself: carnivals always promise another world (Miller 1991).

The rhetorical web

But why start from *words*? Why not start from events, from actors, from situations? We are not saying that there are "only words". We also accept that everyone understands the words for themselves, without a new training; therefore, language does not require a special approach, in one sense. Language matters because people already understand it. Moreover, some approaches to language are forbidding; they look technical and they seem to demand a new training. This book is *not* a linguistics book. Linguistics is a necessary study, the science, of language itself, its structure and functions. Our subject is not language itself, but how environment is discussed, verbally and culturally. Immediately, one acknowledges the inspiration of the wandering linguist, such as John Swales (1990) and Lakoff & Johnson (1980), writers of personal linguistic inquiry.

Indeed, there are many approaches to language. We adopt rhetoric, an approach with a deep history and also a contemporary influence. We believe rhetoric can address the need to think through the effects of words. Today, many fields include rhetoric, and rhetorical approaches are practised in: literature (Fish 1989, Derrida 1992), psychology (Billig 1987), politics (Connolly 1984, Atkinson 1984, Hirschman 1991), law (Perelman 1982, Fish 1989, Goodrich 1986), history (White 1978, Megill & McCloskey 1987), economics (McCloskey 1985, 1990, 1994, Klamer 1988), sociology and social theory (Habermas 1984, 1986, 1987, Schrag 1992), logic (Walton 1989), education (Andrews 1995, Miller 1991), mathematics (Davis & Hersh 1987), philosophy and history of science (Selzer 1993, Campbell 1989, Bazerman 1987, Gross 1989), urban studies (Throgmorton 1993, Fischer & Forrester 1993, Myerson & Rydin 1994). Rhetoric is "decentered", or many-centred (Leff 1987, Miller 1991). No discipline possesses "the" rhetoric. As Miller (ibid.: 39–40) comments, rhetoric is alive because it responds to new situations, it is discontinuous. Leff (1987: 19) remarks that rhetoric is "available for periodic rediscovery". It is not based in one discipline; it is found across many disciplines. And those fields of enquiry use rhetoric; they shape it and are themselves shaped by its insights and usefulness. The rhetorical web is the methodological equivalent of the environet.

Although some textbooks and histories are nostalgically definitive (Vickers 1988), rhetoric no longer has one definition. Instead of asking "what is rhetoric?", it is better to begin by asking: to what does "rhetoric" allude?

> The English word "rhetoric" is derived from Greek *rhetorike*, which apparently came into use in the circle of Socrates in the fifth century . . . attempts were made to describe the features of an effective speech and to teach someone how to plan and deliver one. (Kennedy 1994: 3)

Such passages should not be used as definitions for modern practice. They are suggestive. One can see a link with the usage of "rhetoric" meaning "rant" or "bombast". If oratory is suspect, then "rhetoric" is also dubious. Conversely, if we understand persuasion more generously, then rhetoric is more positive: persuasion can be rational and creative (Billig 1987, 1991, Myerson 1994). To apply today, "rhetoric" must mean more than just speaking well – that refers to the cultural context of ancient Greece (Miller 1991: 39–41). In a modern society, rhetoric is still a living art:

Rhetoric is the art of persuasion, or a study of the means of persuasion available for any given situation. (Burke 1989(1969): 191)

The passage alludes to Aristotle's rhetoric, which studied persuasion, but the tenor is modern. The "art" itself is undefined. Situations vary, and persuasions are diverse. "To persuade" is contrasted implicitly with "to compel"; persuasion can be nice or nasty, elegant or crude, but it is something other than force. It is an art, although not only a fine art. Because no-one can "define" persuasion, because "any given situation" is infinite, there is no defining the "art" of rhetoric. Instead we become spinners of the web of rhetorics, each account adding new threads. Though, as Miller (1991: 39) says, some authorities have supplied "a lament against the fragmentations of rhetoric", ironically only a fragmented art could possibly adjust to the contemporary world. How could one "method" supply an analysis of modern persuasion, let alone give advice on how to persuade?

Even the term "persuasion" is not enough:

Rhetoric is not, however, simply persuasive; it is also constitutive. . . . How we (authors) write and talk shapes who we (a community of authors and readers or speakers and listeners) are and can become. (Throgmorton 1993: 121)

Words make identities, and identities make arguments plausible or weak, charming or repulsive. One can extend the sense of "constitutive": words make worlds for identities to fill out, landscapes of the time within which the argument counts – arguable worlds (Myerson 1992, 1994). The worlds are both general and specialist. Any new rhetoric now faces life beyond "the sharp classical distinction between public and technical discourse" (Leff 1987: 21). The rhetorical question becomes: how are arguable worlds made, and how are resolutions attempted within them?

So "rhetoric" has meant many things, and will mean many more. Rhetoric has meant manuals for orators, technique for poets, theory of knowledge, grammatical analysis, a branch of logic. There are many centres. Among the many centres, we now demonstrate the applications of rhetoric by moving outwards from two contemporaries, each a rhetorical innovator and new traditionalist: Michael Billig and Donald N. McCloskey.

Environment words, environet and the rhetorical web

Michael Billig links rhetoric and psychology, and he also connects ancient culture and modern culture:

> Modern psychology is not as clearly superior to ancient psychology as the sports car is to the chariot. . . . The ancient study of rhetoric can be considered as the forerunner of modern psychology . . . (Billig 1991: 32–3)

The subject of rhetorical psychology is what and how people think, as evidenced by our discussions and dialogues, quarrels and nods, philosophical essays and news broadcasts:

> The project emerges from the development of a "rhetorical approach" to social psychology. This approach, which is critical of much experimental and survey research, emphasises the importance of argumentation. In effect, it says that if one wishes to understand how people think about issues – what their "attitudes" are – one should listen to them discussing." (Billig 1992: viii)

By "argumentation", Billig means a family arguing about the Prince of Wales (ibid. 1992), as well as Richard Rorty arguing abut American philosophy (ibid. 1995). He presents major studies of *Talking of the Royal Family* and *Banal nationalism* (the nationalism of everyday life and of "normal" cultures). He celebrates ordinary argument, the complexity and style of everyday dialogue, the awkward grace of domesticity, and he notices its ideology. Of nationalism, he is both critical and acknowledging: "Analysts must expect to be affected by what should be the object of their study." (ibid.: 37). He uses "flaggings" to refer to a system of nudges, reminders, and influences:

> No course of formal instruction is required to notice the flaggings. Instead, there need be only a conscious willingness to look towards the background or to attend to the little words." (ibid.: 174)

Rhetoric means "noticing", noticing the words and the worlds inherent in them. Most ordinary noticing is lost to academic study. We notice more than we record as analysts or experts. Billig uses rhetoric to prod ordinary attention and to organize these noticings, to provide cultural understanding. Rhetoric cues our attention to things we already notice, and then we notice a bit more too. It gives meaning to noticings that would otherwise seem insignificant. Michael Billig spins a rhetorical web, and in his web we find contemporary conversations and academic texts, presidential speeches and the Talmud. His own language is spontaneous and direct. Rhetoric is not an influx of technical terms; rather, it enriches and alters existing words.

Donald N. McCloskey takes rhetorical stock of economics as a discipline; he also uses rhetoric to discuss the economy. McCloskey is very conscious that fellow economists distrust the "rhetoric of economics", seeing it as an "anti-economics argument":

> The Anti-economics Argument: "Rhetoric is an attack on economics, undermining the claims of economics to be a scientific discipline." No, no. . . . A rhetoric of economics examines all the arguments, and encourages admirable goodness in argument all round. (Klamer & McCloskey 1988: 18)

His reply, with his colleague Arjo Klamer, is that rhetoric notices that economics consists of arguments, it does not satirize economics. Rhetoric is about "all the arguments". As an economist, McCloskey has his views, and they are explicit in his rhetorical work; but as a rhetorician, he is open to all the arguments, to see how they are argued.

McCloskey is analyzing a form of modern knowledge. But is it appropriate to have a "rhetoric" of such specialist knowledge, of science even? Is science not separate from argument, from influence, from persuasion? McCloskey responds:

> The pursuit of Truth is said to be very different from mere persuasion. Yet when set beside the actual behavior of scientists and scholars the notion looks strange. (McCloskey 1985: 46)

So, economics *is* about truth. "Scientists and scholars" do seek truth, they gather facts, they propose theories and test them. The result is knowledge. But the process is persuasion. Economics is persuasive *and* truth-seeking (Habermas 1984). Indeed economics creates ways of writing and talking, "discourses", so that economists can influence each other and also the public: "The range of persuasive discourse in economics is wide, ignored in precept while potent in practice." (McCloskey 1985: 72).

Nothing sinister is implied by "persuasive discourse". Economists believe in their knowledge. They want it to count. Therefore, they try to exert influence. McCloskey defends his rhetoric of economics, and his arguments apply to rhetorics of other modern disciplines, including environmental sciences. There is as much rhetoric in the theory of ecotourism as in a novel on global disaster, as much in the economics of resources as in a disclosure of pollution – and rightly so. Why should the latest facts not be persuasive? They will not speak for themselves. Why should new theories not be articulate? They will not be heard otherwise. Rhetoric, the approach, looks at rhetoric, the language. It asks what the words are doing, why they are lively (or not), why *these* words are chosen to convey the facts and theories. We do not substitute words for knowledge, phrases for ideas: ". . . paying attention to words differs from claiming that economics is not scientific." (ibid. 1994: 324).

The same is true of rhetoric and environmental science. In the quotations on p. 10, we juxtapose the academic journal *Transportation research* and the tabloid newspaper *Today*: but the juxtaposition is interesting because the texts differ, not because they are the same. Rhetoric is about academic discourse and newspapers, specialist articles and policies, working papers and headlines. The texts are different, and some are scientific; but science also argues and persuades. Rhetoric is centrally about creativity. Arguments have created the modern world: "At the root of technological progress is a rhetorical environment that makes it possible for inventors to be heard." (ibid.: 372) Ideas must be heard and knowledge must speak. A language must be available, a language capable of developing facts and theories. If there is to be an environmental revolution, then it will also be a rhetorical revolution. The revolution will be a change of ideas and words: scientific revolutions are argumentative, and cultural change is persuasive (Kuhn 1970, Lindblom 1990).

From Billig and McCloskey, environmental rhetoric can learn certain key principles:
  • Knowledge and ideas are argumentative, as much as beliefs and hopes.

16

- Arguments are verbal, profoundly verbal, although not only verbal.
- Argument is broad, specialist and everyday, technical and public. Rhetoric applies equally. Indeed rhetoric illuminates the arguments for technical status.
- Rhetoric is about creativity, how people make worlds to inhabit, worlds of belief, worlds of knowledge: arguable worlds.

We do not invoke Billig and McCloskey in order to advertise a new (or old) discipline. Rhetoric is not a "second discipline" that must be acquired alongside the modern canon. There *are* specialist arguments about rhetoric's history and theory. But one does not have to train in rhetoric separately.

> Rhetoric was no specialist study, confined to the ambitious few who hoped to make a career from public speaking. On the contrary, it was an established intellectual tradition, which offered practical skills of articulate expression and theoretical insights into the nature of communication. (Billig 1987: 31)

Everyone already knows how to argue about the royal family (Aristotle 1909); and economists already know how to argue about economics. Rhetorical expertise is widespread, but unacknowledged. Rhetoric *acknowledges* our argumentative talents, the words we have invented, the persuasive idioms. Rhetoric's webs become cultural interpretations, interpretations of how facts become relevant, beliefs commonplace, ideas new; and how subjects stay topical – nation or inflation, monarchy or industrialization. Rhetoric draws attention to cultural creativity and linguistic arts: not the arts of special artists, but the arts of "all" who "endeavour to criticize or uphold an argument, to defend themselves or to accuse (Aristotle 1926: 3) – in the present study, the arts of all who write and read (Miller 1991: 39).

In one usage, rhetoric is associated with bombast and fraud: negative views of persuasion. But people do not persuade primarily by shouting and lying. On the contrary: most persuasion is in a lower key, and nothing persuades like the truth, or a good likeness. Billig, McCloskey and others have webbed rhetoric to argument, and argument to truth as much as error, coherence as much as trickery. "Argument" then refers to something fundamental in culture, rather than a kind of quarrel, or an academic or legal technique. The result is an

17

analysis of argument as culture, and culture as argument.

Billig and McCloskey offer a vocabulary whose central terms include: arguing, argument, arguments, argumentative, arguable, argumentation; and also controversy, controversial, disagreement. The vocabulary is familiar, words are enriched not fabricated. But the terms are pulled away from old associations: that is, argument as breakdown or argument as local technique. The title of Billig's book can stand for the rhetorical project of this cultural analysis: *Arguing and thinking*, "the importance of argumentation for thought" (Billig 1987: 48). Rhetoricians inhabit, and interpret "a world of argument and discussion" from which there is no exit:

> Opinions, or individual chains of reasoning, clash in the context of a social argument. If there are always two sides to an issue, then any single opinion, or "individual argument", is actually or potentially controversial. (ibid.: 44)

"Controversy" is universal. It is always possible to imagine another case; one arguer implies another arguer. So argument is "social", and, conversely, being social is argumentative – not *necessarily* antagonistic. If we are thinking, we are arguing, or at least arguable: other voices can be mapped actually in disagreement with us or potentially so. Meaning itself involves differences and alignments. Without the differing, there would also be nothing to agree about. The term "argument" gains in meaning: "Rhetoric is not what is left over after logic and evidence have done their work. . . It is the whole art of argument, from syllogism to sneer." (McCloskey 1994: 35)

Rhetoric aims to show, not correct versus improper arguing, but "the breadth of human argument, the limits to formulas for thinking, the way that words matter to the conclusions drawn, . . ." (ibid.: 396) Rhetoric does classify arguments, but shows how many and diverse they are. As with Billig, "thinking" is the key, arguing as against formulae. There can be no formula because no-one knows what will be said next.

A *new rhetoric*

Our subtitle is "A new rhetoric", echoing other "new rhetorics" such as Perelman & Olbrechts-Tyteca (1969). As Leff (1987: 19) says, rhetoric awaits and invites "periodic rediscovery". "New" does not mean starting from scratch; on the contrary, new rhetoric is implied by previous rhetoric. As the web is spun, new threads spread out. Our rhetoric participates in the webbing of rhetoric/argument/thinking. We present a cultural analysis of environmental texts, and we divide the account into three phases, phases that correspond to classical rhetorical categories. In Chapters Two and Three we present a topic analysis. Chapter Four looks at the ethos of environmental arguments, and Chapters Five and Six consider figures of argument. Each category has links with the rhetorics of Aristotle and Cicero (Kennedy 1994), but each is also contemporary. Throughout, the dominant theme is "invention", to continue the classical theme:

> The first of the five parts of classical rhetoric is "invention" . . .
> This is concerned with thinking out the subject matter: with iden-
> tifying the question at issue . . . and the available means of per-
> suading the audience to accept the speaker's position. (ibid.: 4)

or as McCloskey puts it, invention is "the finding of arguments (1990: 66).

Where do people look for arguments? They look through topics, the arguable issues. The "environment words" are names for topics, rhetorical spaces: resources and energy, population, biodiversity and rainforests; pollution, global warming, sustainable development and sustainability, Earth and Gaia. Other environment words are nested inside: conservation within biodiversity and resources, for instance, and nuclear within energy and pollution. Topic analysis isn't about definition, what the words mean or should mean. The result is not a lexicon. Topic analysis expresses a principle: ". . . the general rhetorical position that different cases can always be made on the same topic." (Billig 1987: 42)

Out of the topics emerge "propositions" each "capable of being opposed argumentatively" (ibid. 1991: 206). Being opposable is not a drawback, on the contrary it is a requirement; that is how we know there is a proposition being put forward. Topics contain contrary voices, and the voices can often be paraphrased as key "sayings":

Environment words, environet and the rhetorical web

The two sets of common-places – "they" should behave better than "us" / "they" are only human like "us" – pose a continual dilemma, which frames each royal action and each royal personage. (ibid. 1992: 96)

Texts find their argument amidst contraries, the topical differences, sayings, and feelings, ideas and intuitions. Ultimately, "topos", the Greek source, also means "place" (ibid. 1995: 96). Monarchy is a topos, or a topic collection; so is environment. But all topics are different. The environmental topics have their own texture, their own density and impact. You find different kinds of argument here. From Aristotle onwards, topics are also the places of knowledge. It is arguments that drive knowledge; in seeking arguments, we also discover truths. Different voices are still possible, but so are insights, discoveries, breakthroughs. The insights enrich the topic, they do not close it down. On the contrary, topics wither away from lack of insight, nothing interesting to say; new knowledge keeps the discussion going, it does not end it.

We have divided each topic analysis into two sections: "Cultural innovation" and "Differences and realignments". Together, they correspond to "inventio": classical invention. Innovation is the search for and discovery of arguments; differences are the motives, the positions that make the search necessary. One implication is that no single "grid" does justice to the environmental agenda in our times. There is no simple polarity: green, anti-green; weak ecological, strong ecological; environmental, developmental. Instead the differences shift, they make new patterns all the time. Partly, that mobility is implicit in argument itself, in the drive to invent, to discover new cases and new things to say; partly, it is a feature of contemporary culture (Giddens 1994).

One term that runs through the topic analysis is "discourse". "Discourse" is a term used in many ways by linguists and cultural analysts. In our rhetorical approach, "discourses" are invented and reinvented to manage the controversies arising from different points of view. We propose three discourses influencing the contested representation of environment in contemporary discussion: new information discourses, new concept discourses and new practice discourses. Through the emphasis on "newness", we link the discourses to argumentative invention.

A *new* rhetoric

New information discourses concern the presentation of information and the discussion of how such information is generated, including the use of statistics, data and quantitative measures (Miller 1994). These frequently involve the claim to "factual" status and the elevation of information into objective knowledge. But the facts do not end the argument: on the contrary, facts fuel the argument. Being arguable includes the disclosure of information, facts, would-be facts; the facts are worth uncovering because of the arguments. And where the existence of uncertainty and doubts about the accuracy of this information are recognized, a sub-discourse of risk and probabilities arises, the discussion of ranges and variability forming part of the information (Sen 1985, 1987, Elster 1989, Rosenhead 1989).

Here is Julian Simon (Myers & Simon 1994: 121) objecting to a new information claim on biodiversity and species loss by his opponent Norman Myers:

Some scientists (in Myers's words) have "hazarded a guess" that the extinction rate "could now have reached" 100 species a year. That is, the estimate is simply conjecture . . .

Myers ripostes by defending the information claim:

What is the evidence for the mass extinction under way? Consider a 1992 book by Professor Edward O. Wilson of Harvard University, *The diversity of life*. Wilson has calculated, using analytic models which are thoroughly established in the biological field, that we are now losing, very roughly reckoned, at least 30,000 species every year. (ibid.: 128)

Simon retaliates: "Garrett Hardin, another biologist, has retreated to the point of saying it is unknowable how many species are being extinguished." (ibid.: 132)

The information discourse is itself arguable. The arguments concern knowledge itself, not just individual facts. This is exemplified by the form that much of the discussion on global warming takes, including the possible consequences such as sea-level rise (F. Pearce 1994a). Information is topical, it generates contrary arguments. Facts do get ratified, authority is achieved; but the authority is never final, rather the discussion moves on.

21

By way of contrast, new concept discourses present new ideas or reinvent old ideas. Whereas a new information discourse surprises and convinces with a new fact, a new concept discourse shocks with a new way of thinking and seeks to persuade the audience into the adoption of the new ideas. The seminal example here is the evolution of arguments around the concepts of sustainability and sustainable development. There is of course a relation between new information and new concept discourses, and many environmental arguments are about this relation. The occasion of argument uses these different discourses to develop the line of thinking and persuasion. Thus, arguments that link sustainable development to the issue of global warming are different from those that link global warming to that other seminal new concept, Gaia.

New practice discourses also develop from reference to new information and concept discourses, but they take the discussion forwards to consider new policies and ideas for environmental management. Here issues of implementation and feasibility are raised along with the question of the relevance of the proposed practice for the intended context. A current example of an issue dominated by new practice discourse concerns pollution taxation, including carbon taxing, and the whole incorporation of conclusions from environmental economics into policy debate.

These discourses are evident in use across the whole range of environment topics. They answer to expectations that are general: How much knowledge is acceptable?; How new can a new idea be?; What makes a proposal practical? The answers proposed to these questions reflect varied institutional bases: disciplines, governmental authorities and agencies, non-governmental organizations, corporations, the press and other media. As such, discourses are social and cultural. They express status and authority. Therefore discourses are worth contesting, especially discourses that are considered scientific. Discourses reflect the operation of influence as well as generating it: UN-sponsored reports resonate differently from NGO documents, which again differ from reports from consultancies or multinational corporations.

Nevertheless, argument is active, and not conditioned by discourses or accepted hierarchies. It is impoverishing to assume in advance that any argument simply reflects power and status, interest and need. Nor can it be presumed that arguments win and lose in line with external power. Winning and losing are not that simple, nor are power and influence. Censorship is real, and budgets do differ; status can buy access. But it does not ensure credibility. Differences and alignments evolve, and they evolve argumentatively, within the discussion. Rhetoric is about content, not about conditioning: it explores how an issue is conveyed, as well as who gets to convey it. That is why we begin with topic analysis. The topics remain an organizing principle when we come to "ethos" and "figures of argument". Ways of arguing are meaningful only in terms of particular issues, contents, and meanings.

"Ethos" is an everyday term. It means atmosphere, moral atmosphere, shared identity. Ethos is a social self, a shared self. It is also about values, the root being "ethical". In rhetoric, "ethos" is about how arguments imply personalities, and how those personalities are value-laden. The key theme is trust, which is also central in contemporary social analysis (Giddens 1992, 1994). Arguments project a persona: what type of person would make this argument? Does the case suggest a trustworthy author? Would you admire or like or believe someone who made this argument? Ethos is not personal, although, of course, the choice of ethos implies personal values and attachments. Ethos addresses current assumptions about people, about who is believable, who grabs attention. It dates arguments: if you read political speeches from other times, it is the ethos that seems most peculiar: how could anyone expect to be believed when they sounded like that? McCloskey (1990: 57) interprets ethos in modern terms:

> Consider the first principle of rhetoric, that the presumed character of the writer affects how the words are read by an audience. The best word is *ethos*, which is to say "habit", "character", "moral impression" . . .

Other analysts have applied ethos to scientific argument: "The ethos established by Gould and Lewontin is, most obviously, that of the openly argumentative critic." (Miller & Halloran 1993: 117)

Ethos is a role. This does not imply something superficial, or inauthentic. On the contrary, effective ethos is likely to be authentic, rooted in the subject at issue, and expressing the commitment of the arguer. In the environet, ethos is about both knowledge and feeling.

For instance, "far-sightedness" is a strong ethos in some environmental discussions; and in others, "reassuring realism" is a good ethos. We also refer, particularly, to the ethos of objectivity, and to "emotive objectivity" and "restrained objectivity". It sounds odd to call objectivity an ethos, an example of ". . . the character that a speaker claims in his speech." (McCloskey 1994: 114). But objectivity is something people express and their texts communicate. It is tied to ways of talking and writing. Of course, the objective claim may be strong or weak: that will depend on more than a choice of words. There must be grounds for objectivity, otherwise the most eloquent expression will not be convincing. Nevertheless, it is, we argue, a mistake to see objectivity as simply a "state of mind" or a method. Objectivity is communicative, and we can recognize the claim verbally. Above all, objectivity is not value-free; on the contrary, it is itself a value (Putnam 1981, 1990), an ethical claim to the benefits and even superiority of the objective stance. Furthermore, being objective does not conflict with being emotive. After all, some subjects imply feeling, and an unfeeling account is likely to seem not objective but pathological. To report a disaster objectively does not mean to avoid feeling; it means to convey the appropriate feeling that goes with a true understanding of the facts. How could an objective view of a tragedy not also be emotive (Nussbaum 1990)? Environmental ethos is rightly about feeling *and* knowing: who would trust an unfeeling argument about famine or extinction?

Neither Simon nor Myers, to re-evoke our two debaters, is unemotive, and properly so:

> The increase in the world's population represents our victory over death. (Myers & Simon 1994: 118)

> . . . food availability per person worldwide has been declining for eight long, lean years. (ibid.: 129)

Both cases imply objectivity, a long strong look at the facts; they also express feelings, exhilaration and sympathetic anxiety. They make a double claim: to know the best information and to feel the right emotion. They ask for a double trust: believe the information, share the emotion. The two claims reinforce one another: you are more likely to share the feeling if you adopt the information, and more likely to

24

believe the information if you empathetically share in the feeling. Knowledge is important, and so is authenticity, in an age when, as we shall argue, knowledge itself is too plentiful to settle anything on its own.

"Figures of speech" is also a common phrase: it's a figure of speech, or "just a figure of speech". The idea is that a culture has phrases ready, or types of phrase ready. "Figurative" also contrasts with "literal" in common usage: speaking figuratively, meaning something more than it says or other than it says. "Figures of speech" have been a problematic element within rhetoric. Rhetoric can become a system of figures, a long catalogue of set phrases and types of expression, immensely complex categories of phrasing. The figurative rhetorics reached a climax in the Renaissance. As Miller (1991: 43) remarks, the tradition is unworkable now:

> Awakened in the dead of the night, few even within the classical school could automatically cite an example of (or perhaps even pronounce) *homoioteleuthon*; few would want to try.

Lanham's "Handbook" tells us that "homoioteleuton" probably means "a similar closing of several sentences or verses" (Lanham 1968: 55). Taken too far, figures are a distraction. They are only useful if they draw attention to something in the content of a text, an argument. Yet, as McCloskey (1985: xvii) says, "Figures of speech are not mere frills. They think for us." He gives an example, and it is the figure of speech that is most often considered, metaphor:

> A rhetorician, however, notes that the "market" is "just" a figure of speech. Yet a serious rhetorician, or a serious philosopher of science, will not add the "just," because metaphor is a serious figure of argument." (ibid. 1994: 42)

Another example makes clear how, for McCloskey, metaphors go deep, they are not just decorations, they involve ways of thinking: "Accounting is the master metaphor of economics. . ." (ibid.: 113). To analyze a new park in terms of "costs" and "benefits" is metaphorical: it sees the community as being like a business, the decision-maker as being like a financial adviser. The metaphor is not sinister, but neither is it neutral . It has effects. It is different from other metaphors, say the metaphor of "the green lungs" of the city, with the city as a body, and the park as its breathing apparatus. Metaphors argue, they are "figures of argument", to adopt McCloskey's phrase, they are ways to "criticize and uphold" (Aristotle 1926).

Metaphorical argumentation is one of our "constitutive figures", along with irony and association. "Constitutive" is adopted from Throgmorton (1993: 121; see p. 13) and adapted to mean aspects of the argument that construct the world of the discussion, the arguable world.

"Metaphor is a device for seeing something *in terms of* something else. It brings out the thisness of a that, or the thatness of a this" (Burke 1989(1969): 247). So metaphor could bring out the spaceship-ness of the Earth, or the tree-ness of nature, the disease-ness of pollution or the greenhouse-ness of climate change. Metaphors make claims, they are propositions, complex proposals. And the same goes for the other constitutive figures.

Irony is also neatly packaged by Burke:

As an overall ironic formula here, and one that has the quality of "inevitability", we could lay it down that "what goes forth as A returns as non-A." (ibid.: 260)

What goes forth as progress returns as environmental catastrophe, in Myers's terms; or what goes forth as ecological concern returns as self-fulfilling catastrophe, in Simon's terms. What looks like a sunny sky turns into a radiation hazard. What seems a clean energy source returns as mass contamination. Irony abounds, so much so that we shall refer to "environmental irony". The world of environmental argument is often ironic, and different sides trade a kind of black humour around the environet. To miss the ironic humour is to miss a crucial ingredient in environmental argument.

The world of environment is arguable; all the more reason to make clear which is the winning argument. The "figures of resolution" are winners, or so they claim: here is the stronger case, the conclusion, the outcome. There is an assumption that "winning" or "losing" is a verdict from outside. But winning is also a performance: texts perform "winning the argument", and to do so they use "figures of resolution". "Perform" does not mean "fake". If you believe you have the stronger case, then why not argue as if you do? Moreover, the figures are not superficial; as McCloskey says, they are also thoughts.

We propose three environmental figures of resolution: "overcoming the polarity", "dialectic of catastrophe" and "the feasible possibility". "Overcoming the polarity" is probably the most influential of all the environmental figures: two elements are in conflict, and yet, having grasped the conflict, we can also reconcile them. Sustainable development has generated some strong examples of overcoming the polarity. The figure proposes itself as a winner, a problem solved, a rival outflanked; its impact depends on the content. "Strong" refers not just to form, but to substance: how deep is the resolution? Here is Julian Simon proposing to win an argument:

> More people, and increased income, cause problems in the short run. Short-run scarcity raises prices. This presents opportunity, and prompts the search for solutions. In a free society, solutions are eventually found. And in the long run the new developments leave us better off than if the problems had not arisen. (Myers & Simon 1994: 123)

Another view is acknowledged: there is a potential contradiction. But then the contradiction is unknotted: "in the very long run, more people almost surely imply more available resources and a higher income for everyone" (ibid.: 124). By contrast, Myers proposes to win using a different figure, a dialectic of catastrophe. The situation is so bad that there must be a solution, soon and also comprehensively: "we are at a watershed in human history . . . environmental breakdown or breakthrough" (ibid.: 125–6).

Two winning arguments, one on each side: the choice is not simple. The viewpoints are not nakedly opposed to one another. They achieve expression in particular words, and specific figures. And the result is a choice, a dilemma involving "the whole art of argument, from syllogism to sneer" (McCloskey 1994: 35).

Discourse and environment

Discourse and environment have been considered by several interpretive approaches. In this section we briefly review these approaches.

27

A rhetorical approach has been applied by Killingsworth & Palmer (1992), in their study of *Ecospeak* in American culture and they have shown that it is possible to look rhetorically at environmental argument. Their approach is related to the present rhetoric by a close focus on texts. But the methods contrast, because Killingsworth & Palmer construct a canon of leading environmental texts as their central focus, arranging others around it. They claim to identify a distinctive, and limiting, language of ecological campaigning and consciousness.

Their key theme is "ecospeak", and they tell its specific story. "Ecospeak" is a language, a language that conditions environmental discussion, or, they argue, has conditioned it. Ecospeak freezes conflicts, it prevents agreement, it leaves environmentalists as outsiders. Both environmentalists and mainstream media have contributed to ecospeak, according to Killingsworth & Palmer. For instance, they use "ethos" to describe an overall environmentalist attitude in debate, an attitude that is disturbing, unable to connect itself with mainstream associations:

> the environmentalist ethos is only now in the process of formation so that the conflicts we feel are part of the inevitable process by which a political stance evolves. In another sense, however, we may worry that as this new ethos solidifies, its characteristic terminology and working rhetoric will cause oppositions to close into an irresolvable set of conflicts . . . (Killingsworth & Palmer 1992: 8)

They fear that the disputes will be prolonged and deepened, that issues will be unresolved, because the presentation is polarizing, instead of unifying. The media contributes to this process, by representing environmental debates between fixed groups, an environmental minority and a mainstream community, the latter including business and commerce: "The conflicts that are reported are irresolvable." (ibid.: 30).

The story is gripping and the analysis vigorous: rhetoric comes to life. They analyze founding texts of modern environmentalism, and their point is to show how, amid the virtues, the dangers of ecospeak are born:

Carson's chief rhetorical strategy was the development of startling contrasts and dramatically rendered conflicts in a future-oriented report on environmental problems. . . . The appeal to emotion in the rhetoric of public debate is always risky. A writer who seeks one response may elicit a contrary one. (ibid.: 65, 71)

Rachel Carson's *Silent spring* is dramatic, vivid and also, they suggest, polarizing.

Elsewhere, science journalism has also advanced ecospeak, using different rhetorical devices:

Ironic Distance. . . . Two nebulously delimited groups – the "environmentalists" and the "public" – are set at odds against one another and are implicitly distinguished from scientific researchers [an article on global warming in *Science*] . . . (ibid.: 149)

But there is hope. Ecospeak seems to freeze the landscape, but new languages evolve:

Until the summer of 1988, when the magazine underwent a major shift in tone and outlook, *Time* remained true to the categories of ecospeak, treating the developmentalist perspective as the dominant viewpoint of the American public and treating reform environmentalism as a protest movement and a minority position. (ibid.: 152)

Different voices can impact. On one side, there is, for example, Greenpeace: "the Greenpeace Foundation has taken a dramatic approach to a rhetoric of resistance . . ." (ibid.: 195). Their style is resistant, abrasive, and still polarizing. A more positive example is the work of Hermann Daly, steady-state economist:

Daly's program has the power of a technical argument, but it departs in so many ways from current economic models that it has the feel of a Utopian discourse. (Killingsworth & Palmer 1992: 243)

Daly is both technical and Utopian. He seems to address different constituencies and he therefore points beyond ecospeak.

Our rhetoric tells different stories, perhaps stories from the latter

29

phase of "ecospeak". We show how the arguments diversify, how the differences shift and realign, how new ideas impact. Two contrasts help to define both our rhetoric and the work of Killingsworth & Palmer. In our view, environmental issues are unresolved for reasons that go beyond ecospeak, although they include it. It is the diversity that renders the arguments unresolvable, not the use of polarizing tactics and over-simple reports of enmities. So we both extend and revise the story: the environet breaks the bounds that Killingsworth & Palmer fear have constrained discussion. But the result is a deeper unresolvability. However, unresolvability is positive as well as negative. All kinds of initiative do emerge, including new knowledge and new truths. It is just that knowledge does not resolve the issues, it alters them.

A second difference is that we do not proceed whole work by whole work. We do not analyze major works individually, nor do we look at major genres individually. Instead, we have created a rhetorical structure: topic, ethos and figures of argument. Across the rhetorical structure, the rhetorical web, we have run environment words. The two are internetted. Our aim is to show where and how individual works find their arguments. Our theme is invention rather than "arrangement", the whole structuring of individual works. There is no such thing as a complete rhetoric: one interpretation rules out alternatives. But we believe individual texts are also illuminated, as they interact across the environet.

Our work implies a dialogue with *Ecospeak*, a dialogic recognition of the initiative. A closer approach, in practice, is that adopted by Burgess 1990; Harrison 1994), an approach that stresses the emergence and interplay of different meanings, addressed to different audiences, or Kempton (1991), also exploring lay and expert perspectives on global environmental issues. As in our approach, the subject of the study is difference, and the stories are multiple. Whatmore & Boucher (1993) also evolve a discourse-based approach that focuses on interaction and the nature of the debate process, in their specific study of planning and environment. Redclift (1994), focusing on the sustainability issue, stresses the emergence of new, or would-be new, discourses on environmental issues, and considers the tensions within them. A critical interpretation of environmental arguments is also developed by Grove-White & Szerszynski (1992), an interpretation that overlaps our approach in its emphasis on the ambiguity of the issues themselves, the ways they elude single definitions. And, drawing on but not limiting herself to anthropological methodology, Milton (1993: 9) brings together a variety of stories that see environmentalism as discourse: "the field of communication through which environmental responsibilities . . . are constituted". Our rhetoric builds on the gathering methodologies that have emphasized such processes and differences within environmental discourse.

The term "rhetoric" also differentiates the approach from "discourse analysis", as exemplified by Bell (1994). Discourse analysis is more closely allied to opinion research (Worcester 1993). The emphasis is on text and opinion in conjunction: "The blending in public discourse which we have discovered between the greenhouse effect and ozone depletion also occurs in media texts" (Bell 1994: 58). Discourse analysis is a more objectivist approach than rhetoric: the categories imply a greater impersonality from the analyst, less room for different interpretations. Discourse analysis seeks to establish a definitive view, whereas we seek to make available different possibilities. Because it is less objective in its claims, rhetoric invites a different dialogue with the reader, and offers an exchange of interpretations and judgements.

In our interpretation, rhetoric is interdisciplinary: it draws on classical rhetoric, modern rhetoric, psychology, economics, literary analysis and philosophy. In linking these fields, the approach crosses over the apparent boundary between the social sciences and humanities. On one side, our concepts derive directly from classical origins, even from Aristotelian rhetoric. But on another side, the concepts relate to contemporary sciences, human and social sciences as they reformulate themselves. The approach is being developed dynamically, in response to the particular texts and ideas concerning environment. We are responding to the environmental arguments, rather than applying a set method to them.

So what?

"The question is, as usual, why it matters." (McCloskey 1985: 185) Our theme is not answers but arguments. We map arguability, and we aim to discover its unexplored extent. To do so, we look closely at words. Anyone who looks at words, and who promotes arguability not answers, will confront "the question, as usual, why it matters". How does it help to show the extent of arguability? Why not leap from words to things? Why not decide who is right? A first reply: an arguable world is a changing and a changeable world. The issues are reshaped, balances alter, and ideas impact. There is a great difference in the nature of the arguable worlds if one view or another view is the major interpreter. There is an arguable world promoted through, for example, the term "global warming", with spaces for different points of view. But there is basic conflict too, and it has practical implications. The arguments matter. To make a viewpoint more persuasive is a real act. Truth is added to the discussion, in the persuasive process of arguing. Measures do emerge; agreements are derived, although also open to reinterpretation later. Arguable worlds are places where things are discovered, conflicts are fought, people agree and differ afresh. Alliances form and dissolve. We aim to explain these processes.

This is a political task. Rhetoric began with democracy:

> Under democracies citizens were expected to participate in political debate, and they were expected to speak on their own behalf in courts of law. A theory of public speaking evolved, which developed an extensive technical vocabulary to describe features of argument, arrangement, style and delivery. (Kennedy 1994: 3)

> Arguable worlds are risky: they may go wrong, and the wrong answers may emerge. But they are also exciting, new voices emerge, ideas spin off, and some of the ideas will also be true. Therefore: "Rhetoric is a theory of democratic pluralism and of general education in a free society. (McCloskey 1994: 385)

"Democratic pluralism" means different things to different theories. It is not everyone's account of the present day. But rhetoric certainly looks for the democratic potential in society, and then the reader must judge how far that potential goes, where it stops and how it might be enhanced and realized. Miller (1991: 39–40) foresees:

> . . . alternative explanations of the relation of rhetoric to modern discourse . . . This new study would have a mandate to replace oral rhetoric with analyses of specifically written discourse – and

with explanations of the inevitably democratic and pluralistic new discourses that result from print technology.

How environment has impacted is a crucial question within the imagined mandate of this new rhetoric of printed democracy.

Rhetoric does not just interpret plurality, providing an understanding of the multiplicity of voices and viewpoints; its understandings are themselves plural:

> If economists tell stories and exercise an ethical sense when telling them, then they had better have as many stories as possible. This is a principled justification of pluralism, an argument for not keeping all one's eggs in a single narrative basket. (McCloskey 1990: 146)

There are more stories and more arguments than can be easily disciplined. On rhetoric's far horizon are other ways of thinking about argument, other responses to plurality than the hope of reduction:

> Nor . . . do I wish intellectual constraints to be loosened. It is harder, not easier, to take into account all arguments in their rich entirety. . . all the arguments, up and down. (ibid. 1994: 317)

> Even contradiction might be welcomed, it might be sought out. For far from being easy, true contradiction can be hard-won, a moment of understanding: "the rhetorical ideal is of *aporia*, where reasonable and justifiable truths come into argumentative opposition with each other, . . ." (Billig 1991: 212)

So the hope of rhetoric is to understand the dynamic and creative processes of argument in a potentially democratic society. It searches for plurality in the communication of that society and it accepts difference and contradiction. It sees scope for rationality in the difference of argument, for the emergence of a truth that is both true and yet also divided. It doesn't offer ready prescriptions for argument:

> A rhetorical analysis has this limit, that it can tell wisely and well how a speech has gone in the past, but cannot be expected to provide The World's Greatest Secret for the future. (McCloskey 1990: 130)

But it does offer a mode of thinking that can foster creativity in argument, in making and reading arguments and thereby maintaining a commitment to questioning the dominant, accepted orthodoxies.

> Yet however attenuated its form, the rhetorical tradition survives to mingle with and trickle through the mainstream of thought. It becomes an anomalous element within the cultural medium. And so it remains available for periodic rediscovery, for direction into a channel of its own, and for use as a corrective to the prevailing drift of mainstream ideology. (Leff 1987: 19)

In this spirit we offer our account of the environet, the arguable world of environment.

Reading the texts

Throughout this and the following chapters we have used quoted material extensively. Such material is an integral part of our text and needs to be read alongside our own words. Some is inserted into the main text, some is separated and boxed for ease of presentation. We have drawn on a wide range of sources and provided an eclectic set of reference points. Our "methodology" for doing so is described in Appendix 1. We note that our references are broader in scope than in many contemporary environmental texts. In part this is a result of our use of information technology (see Appendix 1), but it is also because of an explicitly transdisciplinary frame for our work. We do not propose to create or reinforce the notion of a corpus of key texts; such a corpus might support a discipline of environmental management or environmental politics. Instead we show how "environment" attracts many voices, speaking different idioms; "environment" has the diversity of contemporary culture. The environet is spread across society, including academic communities along with others.

To help the reader bring rhetoric into the reading of the quotations, we have developed a way of annotating the material within boxes. As the rhetorical analysis is extended, so the annotation becomes overlaid and the interpretation enriched. Figure 1.2 provides the key to this annotation.

Discourses	a word$_{NID}$	a word$_{NCD}$	a word$_{NPD}$
	(new information)	(new concept)	(new practice)
Ethos	a word$_E$		
Figure	a word$_A$	a word$_I$	a word$_M$
	(association)	(irony)	(metaphor)

Figure 1.2 Key to the annotation of quotations.

We also encourage others to try a rhetorical approach. To this end we have provided two appendices. The first, already alluded to, describes our method for finding texts across academic, policy and press media. It focuses particularly on the value of new information technology in enabling us to scan across a much larger pool of texts than would have been possible previously. The second offers a sample of texts on the topic of nuclear energy from a variety of sources. We offer no commentary or annotation of these texts but hope that readers will make use of these extracts to develop, individually and in discussion, their own rhetorical analysis.

CHAPTER TWO

Topic analysis (I)

Environet of topics

The environet is composed of topics, and the topics are marked by "environment words". "Environment words" is itself an open catalogue, and a matter of interpretation. But any environment word list would have to include the topic markers of this chapter: resources and energy; population; biodiversity, species, rainforest. The next chapter's markers are as central to the environment catalogue: pollution, global warming, climate change and greenhouse effect; sustainable development and sustainability; Gaia, Earth, planet, globe. Around these words spiral many arguments of environment.

Other words criss-cross every text: nuclear, conservation, ecological, green, environmentalist, industrial . . . Each word marks a cluster of arguments, a topic; and each topic would be a different way to enter the net, to enter from another angle. It is always the same net, but it is viewed differently from whichever topic is chosen. In addition, the arguments are processes. Each topic is recreated by every argument, and so the relations between topics are continuously in motion. The environmental topics are a system, but a changing system. Being a moving system, the net has no starting-point, no original term. Each topic is potentially the beginning. We start with resources and energy, not because it is the only starting-point, but because from here the perspective encompasses some of the greatest arguments on Earth.

37

Topic analysis (I)

Resources, energy

Cultural innovation

Green Mars (Robinson 1994: 101) is a novel about colonizing Mars. As newcomers arrive, they attend a school about resource issues:

> "If you're building a house you can juggle the number of power saws and carpenters, which means they're substitutable, but you can't build it with half the amount of lumber, no matter how many saws or carpenters you have. Try it and you have a house of air. And that's where we live now."
>
> Art shook his head and looked down at his lectern page, which he had filled again. Before clicking on to a new one he read it: "Resources and capital non-substitutable – power saws/ carpenters – lumber-house of air."

The term "resources" hold a key place in the environet. At the heart of the new environmental agenda of the 1990s is the reconceptualization of resources in relation to the environment and the needs of society.

> All physically deterministic models fail to take account of the fact that resources are culturally determined, a product of social choice, technology and the workings of the economic system. (Rees 1985: 35)

This immediately identifies the importance of new concept discourse in discussions around resources. New concepts tend to reorientate the disciplines and redraw the boundaries. In the case of resources the significant boundary is that between economics and the natural sciences and at stake is the status and role of economic discourse.

Economics is classically about the allocation of resources in a context of presumed scarcity:

> Because we live in a world of scarcity, decisions must be made, whether implicitly or explicitly, about how resources should be allocated. The problem of **resource allocation** is solved by the economic system at work in a nation. (Maunder et al. 1991: 75)

And classical economics can claim to provide the key discipline for analyzing and promoting optimal resource allocation:

> Free market capitalist economics is arguably the most powerful tool ever used by civilization. As a system for allocating resources, labor, finance, and taxation, for determining the production, distribution, and consumption of wealth, and for directing decisions about virtually every aspect of our lives together, classical economics reigns supreme. (Gore 1993: 182)

and again:

> There are plenty of ugly images of exploitation of natural resources, and not a few of exploitation of people. But even so this view is flawed. As I researched the resource and human implications of about thirty products and services which were commonly bought in the west but often produced in the Third World, I came to realise that in many cases and places it was the strongly-resourced, widely-experienced multinational that stood a chance of doing things well . . . As a wise economist once said, there is only one thing worse than being exploited by a multinational corporation, and that is not being exploited by one. (North 1995: 294)

This statement of the economic tradition provides a benchmark to which other voices respond, and the responses are diverse. Some sustain the classical paradigms, others propose reforms, remodernizations, even revolutions, all in the context of a robust central tradition, constantly asserting itself in different ways.

The passage from North plays on the different uses of "resource" and "resources" - natural resources, human and financial resources or land, labour and capital. Most classical accounts of economics start from a definition of resources:

> Resources can be defined as the inputs used in the production of those things that we desire. (Mander et al. 1991: 3)

It has to be remembered that by definition resources have no intrinsic worth, but derive value from the useful services or products they yield. (Rees 1985: 37)

But a key aspect of the contemporary new concept discourse is the argument that "land" or natural resources have been treated by classical economics as less constrained than capital and labour, that they are "gifts of nature". Most analytical attention in the past has therefore been focused on problems of allocating labour and capital. Now there is a concerted effort to rethink the significance of scarce or constrained natural resources within the economy. From one way of rethinking:

> Whether they enter the marketplace directly or not, natural resources make important contributions to long-term economic productivity and so are, strictly speaking, economic assets. (Repetto et al. 1992: 365)

or from the Brundtland Report:

> Industry extracts materials from the natural resource base and inserts both products and pollution into the human environment. (WCED 1987: 206)

This reconceptualization involves extending the idea of market failure to apply to the treatment of natural resources, failure arising from the inability of the market to incorporate the full economic value of such resources. This can also extend to an analysis of government failure, of how government policy also works against optimal allocation. Southgate & Whittaker (1994: 21) argue that governments sometimes distort the economy, causing the market to ignore appropriate resources and to over-exploit others:

> By suppressing economic activity in the countryside, distorted macroeconomic and sectoral policies have probably caused some renewable resources to sit idle. But at the same time, those policies have weakened conservation incentives . . .

David Pearce and the London Environmental Economics School are prominent proponents of this position, but this form of new concept discourse is drawn on by many: Lozada (1993) modernizing the model of how interest rates influence resource extraction rates; or Tahvonen & Kuuluvainen (1991) analyzing the conditions for a market solution to the interaction between resources, production and pollution, to take just two examples from a large academic literature. The result is competing claims to technical originality, largely within the discipline of economics and further within neo-classical welfare economics.

But from another direction comes the attempt to propose new concepts in order to revolutionize economic discourse about resources. Daly and others have developed an analysis of the economy in which resources play a fundamental role and, in which the basic nature of all physical matter as energy determines any predictions. Drawing on the laws of thermodynamics, Daly emphasizes that our current economic analyses encourage throughput of resources and ignore any possibility of a limited scale to economic activities. Rather, he argues that the law of entropy means that any rearrangement of resources through production necessarily involves some loss of energy, some dispersal of material, some degree of efficiency of resource use less than 100 per cent: "The market measures the relative scarcity of individual resources; it cannot measure the absolute scarcity of resources in general, of environmental low entropy." (Daly 1992: 189)

Daly produces some vivid ways of reconceptualizing this resource, both a specific resource and the ultimate form of all resources and matter:

> If all of the world's fossil fuels were burned, they would provide only the equivalent of a few weeks of sunlight. The sun is expected to last for another 5 or 6 billion years . . . We must stop talking about free and inexhaustible gifts of nature . . . (ibid.: 22–4)

> Were it not for the entropy law, nothing would ever wear out; we could burn the same gallon of gasoline over and over, and our economic system could be closed with respect to the rest of the natural world. (op. cit.: 16)

The emphasis here is on finitude; the Earth is finite and resources are constantly leaking out, flowing away, never to return in the same form. This Earth is very different from the usual idea of Earth; the stress is on disorder (another meaning of entropy) not unity, on change not equilibrium (the goal of neoclassical economic analysis).

Through the systematic law of entropy, activity creates disorder elsewhere:

> If "elsewhere" happens to be the sun, as it ultimately is for all of nature's technologies, then we need not worry . . . But if "elsewhere" is somewhere else on Earth, as it is for all terrestrial sources of low entropy, then we must be very careful . . . We must stop talking about free and inexhaustible gifts of nature and start talking about the throughput, the entropic flow of matter-energy that is the ultimate cost of maintaining life and wealth. (op. cit.: 24)

Daly is reasserting absolute against relative scarcity. He demands new measuring scales and ways of accounting for resource loss. Daly's economics involves an inevitable limit on the use of resources and imposes a requirement to maintain throughput at levels below a certain threshold. His theory asserts an iron law, modernized by thermodynamics and then modified by innovation. He argues for resource management, in the perspective of ultimate laws of diminution.

Other examples of drawing on new concept discourse are given in Box 2.1. Here Ekins implies that economics needs ethics, value judgements that are overt, views of the good life. On the other hand, Pearce & Turner seek to reform classical paradigms, proposing an interdisciplinary alliance, rather than Daly's transformation of the disciplines. As we have seen, Gore is basically confident: he believes the classical discourse is adequate but he wants to include some new factors, expanding on the environmental dimension of resources.

The texts present new concepts of the discipline of economics itself, rather than simply new concepts within it. But even Daly is advancing an essentially economic discourse, although a reorientated discourse. There are others, however, who present new concepts that challenge both the classical and reformulated economic concepts:

> Ecofeminists have perhaps been most insistent on some version of the world as active subject, not as resource to be mapped and appropriated in bourgeois, Marxist, or masculinist projects. (Haraway 1991: 199)

BOX 2.1   New concept discourse

Theme: the remodernization of economics   Topic: resources

Environmental science and the new concept of economics   "We should . . . recognize that the victory of the West . . . imposes upon us a new and even deeper obligation to address the shortcomings of capitalist economics as it is now practised . . . Many popular textbooks on economic theory fail even to address subjects as basic to our economic choices as pollution or the depletion of natural resources." (Gore 1993: 182–3)

Natural science and the new concept of economics   "The appraisal of natural resource availability and scarcity involves a combination of physical science, materials science/engineering and economic considerations." (Pearce & Turner 1990: 288)

Ethics and the new concept of economics   "Green economics accepts the scarcity of and competition for resources, but insists that economics cannot be value free." (Ekins et al. 1992: 31)

Thermodynamics and the new concept of economics   "Economists claim to study the allocation of scarce resources to different human needs, but thermodynamics is not an educational requirement in the social sciences curriculum. How then could the economists identify "scarcity" of resources?" (Martinez-Alier 1987: 154)

New concepts lead to new practices, which require new institutions. From across the topic, there comes the demand for new practices in the application of economics. One major outcome is green accounting, a new practice that also aims to make new information accessible. The language is that of modernization, of new systems:

The basic ideas of natural resources accounting are:
- to provide an *integrated information system* for the whole resource process . . .
- to measure resources *in physical units* . . . (Lone 1992: 240)

Income accounts for natural resources can be developed directly from accounts expressed in physical units by assigning appropriate monetary values to stock levels and changes. (Repetto et al. 1992: 384)

Ecological accounting seeks to monitor the crucial economic functions of the environment: the resource inputs into the economy; the disposal of wastes from it; and the direct provision of environmental services. (Ekins et al. 1992: 64)

The arguments echo back to a key figure in environmental topics: Malthus, the eighteenth century writer (see below). But the system is new, the idiom is cyber science. The key phrases are: "integrated", "information" and "system", and the key image is that of systematic thinking and practice.

New measurements follow from the new concepts, and new facts emerge in the light of new values. Green accounting is often linked to the attempts to reform rather than revolutionize the concepts of economics, although sometimes it is also linked to the more transformative claims of Daly and others. In its more radical forms, green accounting develops arguments about science and economics, about disciplinarity. One more dramatic proposal has been energy accounting, which implies whole new scales, new concepts, and new professions:

Emphasis on energy accounting was made as a polemical point against economists, who did not study the availability of resources . . . (Martinez-Alier 1987: 154)

The new practices are radical, but, on the other hand, the economic framework is still in place, with its traditional market valuation:

Van Eck manages $1.8 billion. Its primary areas of concentration are global, gold, and natural resource investment. (*The American Banker*, 29 December 1993)

Indeed, most categories of equity mutual funds did quite well. After gold, technology finds were the best specialty funds, up 20.6%. And Natural-Resource funds posted an 18.2% gain, driven by higher US natural-gas prices. (*Business Week*, 27 December 1993)

Natural resource funds are thriving and, at the time of the quote, they were a good investment.

Another way in which new practice discourse is drawn on is in the development of technological solutions to the problems raised by reconceptualizing resources. This particularly applies to the consideration of energy as a key resource:

New ways of thinking about the manufacturing process can lead to large savings in consumption not only of energy but of raw materials as well. (Gore 1993: 329)

And there is a large literature of new ideas and inventions looking at ways of improving energy efficiency and devising means of tapping new energy sources. To take two, not necessarily mainstream, examples: Klaiss & Winter (1992) look at the siting of solar thermal electricity generating installations; and Qu-Geping (1992) considers China's dual-thrust energy strategy. Yet even in these technological discussions, economics does intrude with demands to demonstrate the viability of the different energy options – "Potentially, wind power alone could meet all of the European Community's electricity needs." (Rae 1993: 96) – but what of the costs involved? "The use of wind to generate electrical power is entering a period of massive growth, while the development of expensive nuclear energy is levelling off." (*Herald,* 30 December 1993).

So, Sawyer (1982) discusses the initial high cost barrier in developing new energy systems, the viability of the nuclear option is covered by Beyea (1990), and Rayner (1993) looks at raising energy efficiency as the most cost-effective strategy. Or again:

Fuel cells are not cost competitive with central power generation at this stage of development but they may become so for embedded power generation in heavily loaded urban distribution systems. (Chester 1993: 58)

He said the review should for the first time officially recognise the "vast and unjustifiable costs" of nuclear power, not just to the environment or through decommissioning, but the uneconomic nature of the industry. (*Herald,* 29 December 1993)

This type of new practice discourse also generates its needs for new information. In this case we find substantial amounts of quantitative information relating the costs and returns on investment in different types of energy, notably the comparison between renewable energy and fossil fuels. Walton & Hall (1990) provide reams of data on solar power, for example. The counter-voice is rare; but Coldicutt & Williams (1992) unfashionably do argue for the impossibility of absolutely quantifying solar energy use.

There is a third use of new practice discourse on resources and this concerns the development of a form of environmental management that can conserve or manage resources in line with the new reconceptualizations. The range of views supported here is wide, as will be explored in the next section. And the types of management are also wide-ranging: from the participatory land-use planning advocate by Tan Kim Yong & Fox (1993), to the more technocratic total catchment management of Martin & Lockie (1993), and the integrated natural resource management of Rohlmann (1993).

Energy accounting and ecological economics, thermodynamic resource analysis and new resource measurements, resources as energy and new technologies, issues of viability and new management practices: the new concepts and practices imply a new culture, a culture in which science and economics are at last integrated and the traditional separation between "two cultures" is overcome. For some, like Ekins, the new culture is a revolutionary concept, a cultural revolution; for others, such as Repetto and Gore, the change is more ameliorative. But the debate about resources implies a fundamental debate about culture itself.

Differences and realignments

As discussed above, there is an active reconceptualization of the term "resources" occurring in contemporary culture. This involves a broadening of the concept as well as a reorientation:

> This natural genetic variation – the gene pool – of our flora, and fauna, constitutes the bio-diversity we hear so much about. It is our most precious and fragile natural resource and must be protected. (*Daily Telegraph*, 20 December 1993)

The discussion of resources has therefore become generalized, going beyond the concern with this mineral or that, or with energy resources for human consumption. The alignments between different viewpoints have also become tied to these broader concerns, notably the acceptance (or not) of a broad systematic law concerning resource use and the distributional implications of resource use.

The resource topic is central to the long debate about whether there are limits to growth or not. Although this is inevitably a future-orientated debate, it has long historical roots.

> ... The power of population is indefinitely greater than the power in the Earth to produce subsistence for man ... All other arguments are of slight and subordinate consideration in comparison with this. I see no way by which man can escape from the weight of this law which pervades all animated nature. (Malthus 1970: 71–2)

> Where Man is not, Nature is barren. (Blake 1988: 12)

Malthus is proposing an iron law; nature is sovereign and a sovereign nature imposes laws of limit. Blake disagrees with the iron law viewpoint: for Blake, humanity invents nature, and people themselves produce the creativity of natural world. Nature is only as limitless as human creativity. This is a focused debate with paradigms competing directly. In a Malthusian paradigm, nature is overrun by people, and fights back; in the creative or innovative paradigm, humanity enriches the world; it has neither meaning nor productivity without human beings.

In contemporary culture too, resources arguments are about the relationship between humanity and nature, between iron laws of limits and human creativity, between finitude and endless exploration, between need that cannot be met and the need that motivates people to search, to discover, to progress. The issue of resources, and in particular energy resources, encapsulates the idea that human society is dependent on nature, and the dependence can be constraining. Without nature providing these "natural" resources, we would not, at the most basic level, have heat for cooking and living, let alone the raw materials that underpin our industrialized mode of production and associated life-style. Therefore, nature's limits are humanity's limits. So, there is an argument about constraints. But there is also the argument that people can interact creatively with nature, professionally designated as "environmental resource management".

Many viewpoints develop around these axes. Here the Brundtland
report echoes the Malthusian iron law logic:

> Our report, *Our common future*, is not a prediction of ever
> increasing environmental decay, poverty, and hardship in an
> ever more polluted world among ever decreasing resources . . .
> But the Commission's hope for the future is conditional on deci-
> sive political action now to begin managing environmental
> resources to ensure both sustainable human progress and human
> survival. (WCED 1987: 1)

But the passage still presents the argument of hope and potential. The
debate is present inside the one passage, we can detect the two competing viewpoints. The result is rhetorical ambivalence: resources
mean limits, and resources provide possibilities.

This debate is intrinsic to views about modernization. Since the
late eighteenth century, debate has raged about the way modernity
uses the world. In conditions of emerging modernity, society is no
longer part of nature. Instead, society absorbs nature, and nature is
remade, transformed. The question is then posed: how much
"nature" is there, will modernity run out of nature? Modern societies
are more powerful than pre-modern ones, but they are also more
dependent:

> . . . The power of the Earth to afford a yearly increase of food
> may be compared to a great reservoir of water, supplied by a
> moderate stream. The faster population increases, the more help
> will be got to draw off the water, and consequently an increasing
> quantity will be taken every year. But the sooner, undoubtedly,
> will the reservoir be exhausted, and the streams only remain . . .
> And even this moderate stream will be gradually diminishing.
> (Malthus 1970: 106)

Malthus's argument is about the world as a system and he uses the
water system as his metaphor: the system is running down, the reservoir draining out, not being replenished. It is only human use that
causes the depletion, nature does not run down left to itself. There is a
feedback loop, but by changing the part you change the whole system.
The passage is both literal and metaphorical: water represents all

resources, and water is also an elemental resource. Water is for drink-ing; but it is also, at this time, steam power and water power for indus-try, a form of energy.

There is also the counter-argument against the logic of the iron law. Resources can represent potential, if we interact fairly with nature:

> Conservation has much to do with the welfare of the average man of today. It proposes to secure a continuous and abundant supply of the necessaries of life, which means a reasonable cost of living and business stability. It advocates fairness in the dis-tribution of the benefits which flow from the natural resources. (Pinchot 1901: 81; in Wall 1994: 136)

So, one kind of modernization saps nature, another renews nature. The paradigms of environmentalism, as they emerge from the dis-cussion of resources, are also paradigms of modernization.

The key texts of the wave of environmentalism from the 1960s and early 1970s were similarly centrally concerned with our use of resources. Reference was repeatedly made to the exponential growth in the rate of use of resources that was expected. We would therefore run out of resources by a certain date in the not too distant future. The Malthusian viewpoint remodernized itself in environmentalist language:

> The resources of the biosphere do not make up an unlimited sys-tem whereas the geometrical progression of reproduction seems to be so until famine and death massively intervene. If we prefer, as a species, to employ somewhat less brutal and indiscriminate methods of birth control, then we have to relate population to what the planet can support. (Ward & Dubos 1972: 176)

Malthus provided the geometric/arithmetic progressions, and the images of famine and death; modern idioms added the resources of the biosphere, the predictions of computer models and the explicit concern with system: Malthus's "power of the Earth" becomes the Ward & Dubos "resources of the biosphere". There is an iron law logic but at the same time, Ward & Dubos, like Brundtland, see the potential to manage (in both senses). Either humanity settles for the Malthusian world, or we invent the world of environmental manage-

49

ment of resources. In the rhetorical ambivalence, Ward & Dubos are echoing both sides in the long debate and asking: what is the modern future?

Limits to growth contains another post-Malthusian voice:

> Population cannot grow without food, food production is increased by growth of capital, more capital requires more resources, discarded resources become pollution, pollution interferes with the growth of population and food. (Pestel 1989: 165)

The report of the Club of Rome is often associated with anti-growth arguments. But there is also the drama of modernity, of growing and driving energies. In fact, the writing is also enchanted by the vision of modernity, and the agenda is about safeguarding modernity, not destroying it. Again, we have ambivalence towards modernity, not anti-modernity.

In the event the stories of the 1960s and early 1970s about modernization were modified and transformed by events. The 1973/4 oil crisis following the Arab–Israeli war and the quadrupling of oil prices led to an increase in the efficiency with which oil was being used and a search for alternative fuel sources. There was more conservation and recycling, and efforts in exploration for new sources and alternatives. The industrialized countries altered their production techniques and, more important, their modes of production to emphasize tertiary and quaternary sectors more.

Doomwatch scenarios were altered by the apparent adaptation and the absence of any impending shortage of resources. It seemed as if iron law logics would have to be deferred; the stories of adaptation became more elaborate.

> In the late 1960s and 1970s it was a widely held, much expressed belief that there was an imbalance between the availability of essential resources and future demands for them. A series of books were published presenting an apocalyptic vision of the future of mankind . . . Technological and economic changes have not only prevented the exhaustion of resource stocks, but have done so despite massive increases in population and per capita consumption levels. (Rees 1985: 28, 30)

In the absence of new discoveries, gold, silver and mercury "in the ground" should already be exhausted, and zinc should be mined out by 1990, lead by 1993 and natural gas reserves should be facing exhaustion towards the end of the century. Yet we hear of no such problems . . . The reserve figures in *Limits to growth* are themselves already out of date . . . (Pearce & Turner 1990: 292–3)

This is scene-stealing argument; the key phrase is "no such problems". The logic is *remodernization* as in: "are themselves already out of date". The long debate does not dissolve. The long debate renews itself.

And so to the environment arguments of the 1990s. The Brundtland passage quoted earlier alludes to two visions of the modern world. In the first vision, as society modernizes, the world is "ever more polluted" and people will have "ever decreasing resources"; this is the vision of an iron law of depletion and irreversible usage. The alternative vision is about modernization liberating creativity; modern society is inventive, it makes problems, and then invents remedies. The issue (Habermas 1987, Eckersley 1992) is whether contemporary modernity will work or whether it is self-defeating? Modernity is dynamic (Habermas 1986) and the modern world, like the universe of modern science, is an expanding system. And although resources make growth possible, they also constrain growth. Inherent in the current agenda of modernity, there is the argumentative ambiguity of resources.

Some other contemporary voices in this long debate are given in Box 2.2. Each passage redefines resources, remodernizing the long debate, and creating from it a specifically environmental agenda. In Rees and Brundtland can still be heard Blake's "Without Man, nature is barren": resources are defined by "human ability and need". The Brundtland report is also ambiguous: "people are the ultimate resource". But both definitions run against the view that sees resources as a non-human reservoir that people are draining away; so their definitions are on the human innovation wing of the debate.

Some resource definitions are iron law based, including the definitions by Daly and Porritt, although of course their aim is to redefine the iron law so as to give some scope for innovation and modification. Another style altogether is represented by Haraway, whose ecofeminist redefinition is counter to both iron law definitions and humanist

51

BOX 2.2   Controversies

Theme: postmodernization   Topic: resources

Remodernized Malthusianism   ". . . Exponential growth (in either human numbers or volumes of production and consumption) <u>cannot be sustained</u>_{MPD} indefinitely off a finite resource base." (Porritt 1993: 25)

Remodernized economics   ". . . The upstream industries helping old companies change their processes, conserve their energy, and <u>recycle their "by-products"</u>_{MPD} and former "pollution" and "waste" back into their production stream or find new uses for these unappreciated resources." (Henderson 1993: 103)

Remodernized humanism   "Economists such as Hayek and Robbins . . . Did not <u>study the availability of resources</u>_{MID} . . . Nor did such economists study human needs, refusing to classify them as physiological and ostentatious." (Martinez-Alier 1987: 154)

Remodernized humanism   "It is, therefore, human ability and need which <u>create resource value,</u>_{INCD} not mere physical presence." (Rees 1985: 11)

Remodernized humanism   "People are the <u>ultimate resource.</u>"_{NCD} (WCED 1987: 95)

Remodernized nature philosophy   "Perhaps the world resists being reduced to mere resource because it is – <u>not mother/matter/mutter – but coyote,</u>_{INCD} a figure for the always problematic, always potent tie of meaning and bodies." (Haraway 1991: 201)

Remodernized nature philosophy   "Contacts with the environment are played down because resources are held to be free gifts of nature, <u>not a source of value independent of labour</u>"_{NCD} (Daly 1992: 196)

---

innovation. Haraway defines resources out of existence, and she reinterprets nature from an ecocentric perspective, nature as a living entity, a will in action.

It is notable that these voices show contemporary knowledge as systematic but not unified. There is not one meta-system, no emergent system to synthesize all the rivals. Henderson's system is about rules and aggregated choices, game theory with an ethical heart.

Daly's system orients itself to the physical sciences. The *Limits to growth* system was a precursor of cybernetic translations, society into network. The crucial issue is: which is the predominant ethos of systematic knowledge? What do we trust (Giddens 1990)? The struggle is over knowledge, and who embodies contemporary knowledge.

The arguments are philosophical and practical, they correspond to Nussbaum's (1990) practical rationality. Decisions are being proposed and criticized, but the level of the discussion is also theoretical. Through definition and counter-definition, the long debate focuses and refocuses, and the meaning of nature shifts: nature as energy–matter (Daly) or nature as trickster (Haraway); nature as instrument for humanity (Rees) or nature as sovereign limit (Porritt). Each definition is scene-stealing: to steal the scene from rival economists, outclassed by thermodynamics (Daly) or by scientific philosophy (Martinez-Alier), or to steal the scene from economists altogether (Haraway).

The natural world is a category, and not a stable category. Each definition proposes a different natural world. The whole debate is about remodernizing nature and substituting a modern category for the earlier understandings or misunderstandings. It may sound strange to speak of the modernity of nature (see North 1995). But the debate is precisely about modern nature, and a key aspect is the sense in which nature is a system or systems.

The resources debate has another axis: distributional issues. The question is partly whether there are sufficient resources, but also what are they being used to do, and who is benefiting.

> Climatic factors, geography, the distribution of natural resources operate not on their own, but only within a given economic, social and institutional framework. (Hobsbawm 1968: 37)

From a Marxist perspective, Hobsbawm criticizes fixed laws about resources. Nature is not fixed. Resources are made, and allocated through human action. In effect, Hobsbawm supports the innovative and creative viewpoint, against the iron law view, the creativity being social rather than simply technological.

Social criticism is implicit in the debate about the use of resources:

> The people of the United States today (about 7.5% of the world's population) consume some 30 per cent of the non-renewable

resources produced each year . . . Those who anguish over a starving mankind on the easy assumption that there is just not enough land and resources to go round might do well to try a special kind of visit to their local supermarket . . . How much of the world's land and labour was wasted producing the tobacco, the coffee, the tea . . . How many resources went into transporting and processing them? (Roszak 1972: 404–5)

Roszak is conscious of Malthusian echoes: "a starving mankind", "there is just not enough . . .". But the passage is counter-Malthusian: scarcity in one place is the other side of plenty elsewhere. Like Hobsbawm, Roszak extracts issues of equity, issues potentially suppressed by the logic of the iron law. And these are issues that can be contained by emphasis on a common denominator, valuation in money.

Our uplands and coasts are an integral part of our national identity and, if properly managed, are an ongoing natural resource of considerable value to the whole country, not the sole possession of private individuals or State agencies, but an asset held in trust for future generations. (*Irish Times*, 17 December 1993)

Therefore, it is the debates around the iron law and the practice of common valuation practices that dominate much of the discussion. The long debate is renewing itself through the topic of resources and energy. And in doing so it is renewing the unresolvable debate on the future of modernity.

Population

Cultural innovation

Population is a rich topic, flooded with information claims and charged with intensity. Many presuppositions about arguing need to be revised if we are to understand the dynamics of population issues: presuppositions about what constitutes debate, and how debates are "won"; presuppositions about the role of knowledge in debate; and the presupposition that knowledge is separate from feeling.

For emotion and information are integrally connected in the population question. Beneath the detail, the question is one of imagination: how to imagine the world, the world as a global mass of humanity. Therefore, we find a philosophical dimension in the discussions, asking what kind of concepts structure our feeling about the world? But this is not a new concept discourse, it involves a redefinition, a reinterpretation of concepts – a competitive reinterpretation of world concepts.

As Hardin has said: "The complex of concerns we blanket with the name "the population problem" has been with us for almost two hundred years" (Hardin 1991: 162). Or as Bookchin puts it most vividly:

> The "population problem" has a Phoenix-like existence: it rises from the ashes at least every generation and sometimes every decade or so. The prophecies are usually the same – namely, that human beings are populating the Earth in "unprecedented numbers" and "devouring" its resources like a locust plague. (Bookchin 1994: 30)

So, arguments about population involve reinvented concepts. And yet these are not "old arguments", for they are redolent with contemporary emotion based on new information. The new information discourse on population constantly presents us with new facts, often overwhelming in their scale. The texts claim status and knowledge through such material.

But a feature of the use of new information discourses in the population debate is their intensity. Since classical times, rhetoric has identified intensification and amplification as key features of argumentative discourses on topics; and intensified and amplified new information discourses are key to the topic of the population. Population arguments are full of facts, claimed and apparently proven facts: the population size, the rate of growth, the scale of scarcity, the number of years left . . . Yet the arguments are also intense, heated and emotional. The character of population discourses as intensified and amplified, personal and visionary, need not conflict with the presentation of factual material. Rather, the intensified tone adds conviction to such factual presentation.

Intrinsically amplified information claims are also involved in the counter-arguments about the factual basis of the population "problem" (such as Dyson 1995). A key exponent of the emotive counter-argument to the overpopulation stance is Murray Bookchin:

> when we are told that the "population issue" is merely a "matter of numbers," as one zero population growth writer put it, then the vast complexity of population growth and diminution is reduced to a mere numbers game, like the fluctuations of Dow Jones stock-market averages. (Bookchin 1994: 32)

His position is rather that: "Under such a [capitalist] society the biosphere will eventually be destroyed whether five billion or fifty million people live on the planet." (op. cit.: 34)

Another feature of the new information discourses used relates not just to the presentation of factual material but to the emphasis placed on the process by which such material is generated (e.g. ODA 1991). The discourse is methodical and methodological. There is frequent emphasis on the "scientific" process of producing facts to support claims to knowledge. Hence, there is the extended discussion of the use of modelling, an activity that underpins the whole discipline of demographic studies and its claims, particularly its predictive claims. Modelling is also a characteristic feature of the discussion of population–environment relations. These points are illustrated in Box 2.3.

The second quote from NAS illustrates a key theme – the way in which environmental degradation is generated by population pressures (see also McCormick 1992 or Inman 1993). This often involves quantification, seeing environmental damage in terms of numbers and the interaction between humanity and the environment in terms of statistical relationships, often arising from quite sophisticated modelling (Lutz & Baguart 1992).

An influential example is the Ehrlich & Ehrlich (1990) equation, $I = PAT$, where I is environmental impact, P is population, A is per capita consumption and T is environmentally harmful technology, the three independent variables acting together multiplicatively. Another variant is provided by Southwood:

> Human impact on the environment (I) is basically related to three factors: population size or the number of people (P), the amount of energy used per capita (E), and the extent that the energy used is non-renewable or leads to non-reversible degradation of the envi-

---

BOX 2.3   New information discourse

Theme: the problem of numbers   Topic: population

Too many people in the space   "A finite world can only support a finite population; therefore, population growth must eventually equal zero." (Hardin 1968: 289)

Too many people in the space   "A discussion of the human causes of global environmental change would not be complete without mention of the pressures that the sheer numbers of people inhabiting the Earth place on the environment." (NAS 1990: 50)

Too many people; the contraction of timescales   "Whereas it had taken mankind more than a million years to reach a population of 1 billion, the second billion required only 120 years; the third billion, 32 years; and the fourth billion, 15 years". (McNamara 1991: 49)

Too many people, apparently complicated by uses of resources   "It is quite clear that the more numerous poorer countries use much less energy per capita . . . ; one might argue simplistically that at present levels of energy use, every child born in the USA has 72 times the impact on the environment of a baby in India or 200 times that of the hunter/gatherer child." (Southwood 1992: 8)

Too many people but relative to economic growth   "No simple relationship links population levels and the resource base . . . Whereas a government might be fully capable of providing food, housing, jobs and health care for a population growing at 1 per cent per year (and therefore doubling in 72 years), it might be completely overwhelmed by an annual growth rate of 3 per cent, which would double the population in 24 years." (Tuchman Matthews 1993: 29)

---

ronment (N). These relationships may be simply expressed as . . .

$$I = (P \times E) + (P \times E \times N) \text{ (Southwood 1992: 6)}$$

And yet another variant is provided by Myers's use of catastrophe theory to suggest precipitous environmental decline once population reaches a certain scale: "A situation that seemed as if it could persist into the indefinite future suddenly moves on to an altogether different status" (1992: 21).

Underpinning this approach to representing the dynamics of population change are some persuasively technical terms, on which policy can then be built. One such term is "the demographic transition", often expressed graphically. This is used to show how the situation in the North differs from that in the South, and it can even be posited as an explanatory tool – with industrialization and development will come reductions in population growth implying certain development paths as the preferred population policy (e.g. Bradshaw & Fraser 1990, Wortham 1993).

Another term in population methods is "carrying capacity". This can be defined as the amount of human load that an ecosystem can sustain without causing social, economic and political problems. The use of this term can be controversial, given the tendency to assume that, once identified, carrying capacity is fixed and immutable instead of determined by and subject to societal change. Catton (1993) uses this definition to explore two different explanations for the extinction of the Easter Island culture; in the first, the carrying capacity ceiling was seen as lying well above possible population increases, so that alternative causes, such as a geological catastrophe, had to be invoked

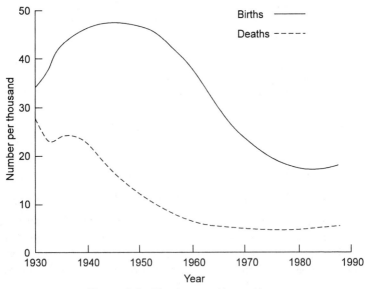

Figure 2.1 The demographic transition.

to explain the death of the culture; in the second, the population levels on the island were seen to have exceeded the maximum sustainable load, inflicting environmental damage that reduced carrying capacity and causing the self-destruction of the culture. Thus, through this technical term, the environmental consequences of general population change are given specific and explicit expression.

This discussion of the population issue as one of new information, of trends and predictions, spurs the use of new practice discourses. New practice discourses go beyond the consideration of information to answer the question "What should be done?". This is largely predicated on the assumption that something must be done, that we must act now, before it is too late. The proposed schemes are diverse with a scientific, technological or sociological justification; for example, contraceptive provision, systems of state subsidies, programmes of sterilization, and increased education for women. Indeed, the new practice discourse is a more contested terrain, bringing out the different viewpoints (see below) more sharply than the associated new information discourse.

The tones are almost always intense, urgent, demanding, even desperate. The aim is to counter the prevailing view that "Providence" is the root cause (Hardin 1991: 162), to prevent fatalism, but the tone is also because of the difficult decisions that are posed:

> It is fair to say that most people who anguish over the population problem are trying to find a way to avoid the evils of over-population without relinquishing any of the privileges they now enjoy. They think that farming the seas or developing new strains of wheat will solve the problem – technologically. I try to show here that the solution they seek cannot be found. (Hardin 1968: 289)

And then there are, of course, vested interests in promoting such a view:

> The problems which appear to arise simply from over-population were the major motivation for the creation of what has been called the "development industry", the business of scholarly analysis and the provision of advice to both donors and recipients of international aid. (Young 1990: 109)

Where the argument is put in this way, the result can be an authoritarian approach to population control practices (e.g. Shaomin-Li 1989, and as critiqued in *The Ecologist* 1993: 145).

Population discussions are used to present more general positions on how the world is changing; population trends are seen as an indicator of the state of the world, and as a symbol of the future. The new information and new practice discourses involved are, therefore, carriers for broader social viewpoints. However, the nature of the arguments means that the full range of such viewpoints is not given equal emphasis. This is explored in the next section.

Differences and realignments

Three key issues, each surrounded by a cluster of viewpoints, can be discerned in the arguments on population. These concern the extent to which population growth is a problem, the question of whose problem it is, and the nature of the relationship with the environment.

As Smyth (1995: 3) has pointed out, the integration of population and environmental agendas has confirmed the definition of the population issue as the "problem" of overpopulation. Hence, there is an emphasis on the scale of the problem, on the scale of population levels and growth rates. This is generally presented in massive, even catastrophic, terms, posing the ultimate "challenge" to politicians and policy-makers. Thus, Meier-Braun (1992) places the "population explosion" alongside problems of war, infringement of civil liberties, North–South economic disparities and environmental catastrophes, and Le-Houerou (1990) argues that the impact of global warming would be trivial by comparison with continuing population growth. Comprehensively, Farrenkopf (1993) links the "population explosion" to the proliferation of weapons of mass destruction, the intensifying global ecological crisis, the global economic slow-down, the deepening North/South divide and the persistence of cultural, ethnic and nationalistic conflict to make a claim for the "illuminating quality of historical pessimism". Or as Gardner puts it (1992: 19–20):

> Where in the range of 7.8 to 28 billion population does finally stabilize will fundamentally determine the prospects of the human race not only for a habitable planet but for human rights, political stability, and world peace . . . Can anyone doubt that if even these medium growth figures are realized, our children

and grandchildren will witness unprecedented misery, world-wide violence, and a tidal wave of unwanted immigration coming their way?

It is of course possible to consider a nation's population as a resource, a source of labour and an extensive market, rather than as a burden of dependency and a user of resources (ODA 1991, Simon in Myers & Simon 1994). In Box 2.4, the selection of quotes from the newspaper media illustrates these alternative positions – population growth underpinning crowded prosperity or generating poverty. However, positive views of a growing population are not as well represented in the popular, policy and academic debates as negative views. This relates to the second issue.

This second key issue is the question of whose problem (or less commonly resource) is represented by the population under discussion. Given the nation-state level at which population figures are presented, this becomes a question of how individual countries or groups of countries view their own population and their attitude to other countries' populations. In part this *is* a statistical question; demographic change clearly takes very different forms in the older industrialized countries, where stability and ageing are the norm, and the developing countries, where the major growth in numbers is found.

But it is also a matter of the values and ideologies of individual societies and the nature of the relationship between different countries. The 1994 International Conference on Population and Development, held in Cairo, highlighted both the extent to which certain developing countries resisted the prevailing definition of their growing populations as a problem and the resentment felt by developing countries that the "population growth = problem" equation was one being drawn by the developed countries.

Rhetorically, the significant aspect of the different viewpoints is the extent to which these regional variations on a global scale are wrapped up in a common problem, a global problem being by definition a common problem. This globalization is often used to override local variations in the perception of a problem, as well as to favour particular solutions. Its use by the "population establishment", the key international population organizations, is therefore significant. Here are two examples, from Shridath Ramphal (quoted in Gardner 1992: 49):

Topic analysis (I)

BOX 2.4   Controversies

Theme: population and the future   *Topic: media bite on population*

The prosperous crowd, the prosperous future

INDIA   "American businesses are flocking to India . . . 'There is an educated, urban middle class that is larger than the population of the entire United States,'ᴹᴵᴰ says McDonald's Corp spokesman Brad Trask. 'That's a very impressive market.'" (*Washington Post*, 31 December 1993)

CHINA   "Despite all the concerns over NAFTA, China's fast-growing economy and huge population continue to attract tens of billions of dollarsᴹᴵᴰ in foreign investment." (Dow Jones, 30 December 1993)

JAPAN   "In Japan, population density is 133 per 100 acres,ᴹᴵᴰ compared with 11 in the United States . . . Per capita, Tokyo residents have only about one eighth as much parkland as New Yorkers . . . It's a crowded country . . . The country has one of the world's longest life expectancies . . . Japan has 274,000 industrial robots,ᴹᴵᴰ compared with 41,304 in the United States . . ." (*Associated Press*, 30 December 1993)

The destitute crowd, the impoverished future

INDIA   "The Asian subcontinent is enjoying a period of political stability that may let its feuding nations work on their endemic problems of poverty, ethnic conflict and population growth." (*Associated Press*, 31 December 1993)

BRAZIL   "3% growth is barely enough to absorb the growth of the economically active population, still less to tackle the backlog of social deficiencies that has accumulated over many years." (*The Economist*, 25 December 1993)

BRITAIN   "How to fund state pensions against the background of an ageing population has become a major worry. There are currently 3.3 peopleᴹᴵᴰ of working age for every pensioner but the ratio would have fallen to 2.2 by the year 2030ᴹᴵᴰ but for the decision to raise the retirement age for women . . ." (*Daily Telegraph*, 27 December 1993)

Each of us – man, woman and child, rich and poor, of whatever faith, whatever race, whatever religion – must begin to take up our mutual dual citizenship. We must all of us belong, and have a sense of belonging, to two countries – our own and the planet.

62

and from Robert McNamara (1991: 48):

> Developed and developing countries have a common interest in
> avoiding the consequences of current population trends . . .
> Unless such action is initiated, the penalties to the poor of the
> world, individuals and nations alike, will be enormous. And the
> ripple effects – political, economic and moral – will inevitably
> extend to the rich as well.

In the following example, though, we can see scepticism towards this
global conception of the problem:

> Few people doubt that population pressure exists in particular
> areas, or that local population crises, perhaps very severe ones,
> may develop . . . The argument is inherently difficult to make
> for the world as a whole, however, because conditions in differ-
> ent parts of the world are so divergent. (Feeney 1991: 74)

Another aspect of the attribution of the population problem has
been highlighted by feminists. The Cairo Conference emphasized the
role of a policy approach focused on women in "solving" the population problem, but, as Smyth (1995) analyzes, there are different
feminist perspectives on the population issue. The mainstream position locates both gender and population within the context of development concerns, and highlights the significance of international
power imbalances in relation to all three issues. The radical feminist
perspective places more emphasis on the social relations of patriar-
chy, on the exploitative use of science and technology and, in its
ecological feminist manifestation, adopts a "naturist" approach to
women. At issue in these disagreements between the dominant views
of the overpopulation problem, the mainstream feminists and the
radical feminists is the extent to which policies aimed at women's
health, education and employment are primarily concerned with
population control and not women's continued exploitation.

The third key issue is the nature of population–environment interactions. In parallel with the often dominant viewpoint that population growth is a problem and is a global problem, there is the
assumption adopted that population growth is automatically an environmental issue (ODA 1991: §1.1). This is usually supported, of

course, by material that more closely examines the nature of the relationship between population growth and environmental degradation. These materials can be anecdotal, drawing on "common sense" logic as typified by Garret Hardin, one of the key commentators on population issues:

> The pollution problem is a consequence of population. It did not much matter how a lonely American frontiersman disposed of his waste. "Flowing water purifies itself every ten miles," my grandfather used to say, and the myth was near enough to the truth when he was a boy, for there were not too many people. (Hardin 1968: 293)

Or there can be more specific and sophisticated analyses, as in the papers in Davis & Berstam (1990), Tolba & Biswas (1991) and van Imhoff et al. (1992).

However, as with the development of methods to generate predictions of population trends, these analyses of the population–environment interactions do not settle the arguments. Rather, they serve to redefine and restate the "problem". The ensuing arguments are then characterized by emotion and intensity and a recognition, explicit or implicit, of the profound distributional issues at the heart of population discussions. The different sides of the arguments, the pro and con, do not interact and they cannot, therefore, contest each other, only displace each other. So, the effect is a heated contest, a contest in which it is not always possible to identify the speaker from the argumentative stance. Imagine a radio report of arguments against the widespread use of contraception – what would be the gender, the nationality, the political persuasion of the arguer?

At the same time the different arguers use the population issue to project different worlds. Population is a carrier for broader philosophical positions. This reinforces the polarization of views. Because there are so many polarities, the overall form of population discussions is not a unified debate but rather an antagonistic cacophony. There is a network of conflicts, but no overall contest. Therefore, despite the predominance of one particular viewpoint, no side can "win". There is no "totality", no grand antithesis and, therefore, there is no grand advance, no single winning argument. There can only be local victories on the population issue.

Biodiversity, rainforest, species

Cultural innovation

The term "biodiversity" applies to the variety and variability of living organisms from genes to elephants. (Grant 1995: 66)

> Since the Rio summit, the world has lost an estimated 40,000 to 80,000 species. (Pearce 1994b: 5)

> One aspect of the process of changing government and popular perceptions about biological resources is to show that the sustainable use of biodiversity has positive economic value, . . . (Pearce & Moran 1994: 15)

Biodiversity is now an established focus for academic and professional attention. Ehrlich & Wilson (1991: 758) refer to "the rise of biodiversity studies: the systematic examination of the full array of organisms . . . together with the methods by which diversity can be maintained and used for the benefit of humanity". But this is not a purely professionalized discourse. There are many related controversies concerning property rights, indigenous rights, North–South relations; the topic of the tropical rainforests captures many of these. And there are also links to animal rights and the relationship of the "human" to the "natural".

This gathering of "environment words" around the biodiversity issue reflects the active use of a new concept discourse. But biodiversity discussions use a new concept discourse of a distinct kind. Biodiversity is a cultural inheritor. "From Linnaeus to Darwin to the present era of cladograms and molecular evolution, a central theme of biology has always been the diversity of life." (Ehrlich & Wilson 1991: 758) Behind the debates around biodiversity is a grand tradition: the evolutionary paradigm in science and in culture.

Evolution is of course a central scientific concept, a connecting thread from studies of organic chemistry and DNA (Dawkins 1982) to understanding ecosystems at different levels. But at a cultural level, evolutionary thought also satisfies a need for concepts to imagine life. Through evolution, biology has shone images of life into modern culture; those images are now illuminating contemporary discussions of biodiversity and, dialectically, biodiversity debates are forming discussions about life. In these discussions we are testing the concepts to see how far Victorian biology will stretch to meet the cultural needs of the twenty-first century. Will the surviving great modern theory of evolution also be the first postmodern theory?

So, the new concept discourse used in relation to biodiversity involves an active reformulation and reinvention of Darwinism: ". . . Biodiversity is a very complex and all-embracing concept, which can be interpreted and analysed on a number of levels and scales." (Pearce & Moran 1994: 6). Or, as Thompson & Welsh note (1993: 32) "The maintenance of biodiversity is an important conservation issue, yet what constitutes biodiversity is not well-understood"; they refer to the IUCN/UNEP/WWF definition as: "the total variety of genetic strains, species and ecosystems", which is continually changing, evolution giving rise to new species and ecological conditions causing others to disappear.

And there are networks of concepts; the biodiversity concept is strongly related to the subsidiary concept of species. Darwin's arguments began from argumentation by definition: he defined "species as only strongly-marked and well-defined varieties", not "a special act of creation" (1968: 110). For some contemporary eco-philosophers, species are genetic, and also ethical:

A species, an ongoing genetic lineage sequentially embodied in different organisms, is evidently a living system . . . From a more biological point of view, a species has an interest in maintaining itself in functional equilibrium with its environment as a matter of its continuing viability. (Johnson 1991: 156–8)

The introduction of another concept does not reduce argument nor resolve them; on the contrary "species" is also a contested term. Gould (1989: 42–3) questions our assumptions about species and how they are interrelated: "Nature's theme is diversity . . . Why, then, do we usually choose to construct a ranking of implied worth (by assuming complexity, or relative nearness to humans, for example)?" "Species" is an arguable term, because evolutionary theory is arguable; biodiversity, therefore, links to an arguable world.

The contested new concept discourse meshes tightly with the new information discourses on biodiversity. Currently:

The debate over the extent of diversity loss is befogged by a lack of information. No one has a clear idea of how many species have been lost; and no one has a clear idea of what is now extant. (Vellvé 1993: 64)

But there is also an information "big bang", the universe of biodiversity facts keeps expanding on all sides. Within this there is a strong focus around the question of measurement and specifically on methods of measurement. For example, McKendry & Machlis (1991: 139–40) adumbrate biodiversity "gap analysis", using "maps of estimated biodiversity", whereas Albert (1993) uses a geographical method based on "species densities".

This emphasis on measurement provokes a cultural irony: it is the measurement that makes the concept elastic enough to be open to interpretation from different perspectives. Quantification here fosters argument and does not close it, even though (or perhaps because?):

. . . in an early stage, and surprisingly so, is the elementary taxonomic description of the world biota. (Ehrlich & Wilson 1991: 758),

and:

On our planet, knowledge of biota is limited and scanty. For instance, there are only 3000 scientists who specialize in tropical biology and 1500 taxonomists who are devoted to the tropical areas (Paoletti et al. 1992: 6)

As the new information discourse expands and the discussion of methods becomes more complex, there is also more emphasis on uncertainty, on reflexivity, on the way that methods alter facts and on the immanence of values in the data.

One key problem is to decide "exactly what we measure in order to assess biological diversity" (Pearce & Moran 1994: 6). Hammer et al. (1993: 97) point out: "The concept of biodiversity is wide and rather complex, sometimes causing confusion of its meaning." They use the distinction between species diversity, genetic diversity, functional diversity and spatial and temporal diversity. Complete measurement is clearly not possible:

At present, probably less than 5 per cent of the biological diversity of the rainforests is known to science (McNeely et al. 1990). The task of recording all of the Earth's life forms would probably take the entire working lives of 25,000 taxonomists. (Schücking & Anderson 1991: 15)

So, what do we select? This is not just a question of techniques of scientific measurement. Measuring biodiversity involves choices of imagination; what do we include in the measurement of life? Physics projects images, images of the universe in space–time, images that reach us through "popular science". Biodiversity is like space–time: it provides a new image of nature, and of nature's scale and complexity.

Although 150 plants are commonly eaten worldwide, only 15 plant species provide more than 90% of world food . . . And only three (rice, corn and wheat) produce almost two-thirds of this amount (Paoletti et al. 1992: 9)

Australia is one of 12 megadiverse (extensive total biodiversity) countries that together account for 75% of the total biodiversity of the planet. (Mott & Bridgewater 1992: 284)

So, biodiversity is about measurement, and it is also imaginative. These measurements do not resolve the conversation since that measurement involves prior decisions and definitions: what is being measured, and what is the scale? And the definition involves an attempt to imagine the scope of biodiversity.

No one definition has as yet satisfied all naturalists; yet every naturalist knows vaguely what he means when he speaks of a species. (Darwin 1968: 101)

The renewed concept and new information discourses of biodiversity are, therefore, closely interrelated, with the act of measurement uniting fact and value (Putnam 1981, 1987, 1995). This could be seen to create a problem: how to deal simultaneously with the concepts and the information. On the other hand, it creates an opening, a chance for cultural change: biodiversity could be culturally central since it is so concerned with the interface between informational and conceptual issues.

But this is not just a debate where concept and information interact in scientific terms. The science is called upon to meet other demands, urgent practical demands to solve problems. The biodiversity agenda is based on a new practice discourse also. Indeed, the new concept, information and practice discourses are unusually closely related in the case of biodiversity.

The use of new practice discourses ranges from the global to the local. At the international level, biodiversity as a topic has received official recognition and sanction; as with global warming, it is the official face of environmentalism:

> Biodiversity and its conservation will be high on the agenda for world leaders next month when they gather in Rio de Janeiro for the Earth Summit. Some will be thinking of as-yet undiscovered medical riches, waiting in the rainforest. A few might be aware of the diverse ecological services such intact habitats provide, such as cleaning up pollution . . . (*New Scientist*, 9 May 1992: 112)

> Many of those involved were cautious about the prospects for successful negotiation of a biodiversity convention given the limited time frame, and the size and complexity of the task. Given that, it was a remarkable achievement for governments to arrive at consensus on the need for a more global and comprehensive approach to conserving, and using, biological resources . . . (Grubb et al. 1993: 82)

At this level, biodiversity links global politics to science, by way of quantitative evaluation. Problems arise, of course:

> The CSD [UN Commission on Sustainable Development] is trying to reach agreement on a new UN Forest Convention. But its work is being co-ordinated by Canada and Malaysia, two leading timber producing nations heavily criticised by conservationists for their destruction of natural forest biodiversity . . . (Pearce 1994b: 5)

The politically achievable consensus is limited, but biodiversity is not undermined as a concept. Its scientific nature is not denied by legis-

---

BOX 2.5   Controversies

Theme: the management of nature   Topic: biodiversity

Unification through management   "The conservation of biodiversity must not
fracture into an endless series of independent skirmishes over research priorities. The main thrust will have to be management: how society applies its
knowledge and determines what is needs to know, and how changes in human
behaviour and institutions can be achieved. The failure to put adequate management in place is a failure to face problems and opportunities". (Mott &
Bridgewater 1992: 287)

Sensitive management   "A wise management of the area would not break
down the delicate balance of the nutrient cycle." (Prance 1990: 895)

Moderate management   "It is unfortunate that extreme applications of management systems that result in high biomass yields have led to a general repu-
diation of management activities with regards to the conservation of biodiversity.
We propose that management activities, including the use of monoculture tree
plantations, can be useful for restoring the species composition of damaged
sites." (Lugo et al. 1993: 106)

The management of preferences   "Indeed, the valuation of preferences for
biodiversity is perhaps the most challenging issue in the context of economic
valuation." (Pearce & Moran 1994: 48)

---

lative problems. So, in the article above, the critique comes from
within the concept itself; it is not the concept that has failed, it is
the agreement that is limited. This can create a central tension in biodiversity discussions between official politics and scientific inquiry.
The politics needs answers but the science has become more intro-
spective about its criteria, more critical of its own limits.

At the local level too, measurement is the key to the new practice
discourse. New styles of management are advocated, in response to
new styles of measurement (Box 2.5). Del Amo & Ramos (1993) propose "diversified units" in forest management. The aim is to reconcile
"production and biodiversity". Thomas & Kevan (1993) contrast
diversity to "vulnerable monocultures" and they advertise new agri-
cultures, against the orthodoxies of modernization. Watkins et al.

BOX 2.6   New practice discourse

Theme: managed worlds   Topic: rainforests

Social management   "Deforestation in tropical areas raises serious health problems which can only be resolved by sound forest management, settlement planning, adequate infrastructure and health care." (Independent Commission . . . 1986: 21)

Spatial management   "Unless appropriate management measures are taken over the longer term, at least one-quarter, possibly one-third,[NID] and conceivably a still larger share of species existing today could be lost. Many experts suggest that at least 20 per cent of tropical forests should be protected . . ." (WCED 1987: 152)

Production management   "If a balance between the sustainable use systems and the conservation of natural areas can be reached there could be a future for the Amazon forest." (Prance 1990: 900)

Local control   "Indigenous peoples need to exercise their control over the resources of the rainforest . . . (Gray 1991: 73)

Strategies of non-intervention   "Given that the industrial sector does not benefit from the diversity of species and uses of trees, forestry programmes deliberately destroy diversity for increasing yields of industrial raw material . . . Naturally diverse tropical forests are considered "unproductive" . . . Productivity of monocultures is low in the context of diverse outputs and needs." (Shiva et al. 1991: 48–50)

(1993) recommend eco-tourism, carefully managed in relation to biodiversity, whereas Mitchell & Barborak (1991) recommend coastal parks and Thompson & Welsh (1993) propose "integrated resource management". Further examples of the new practice discourse are given in Box 2.6, drawing specifically on the case of rainforests.

Each example starts from measurements, of species loss or habitat loss. Some responses are managerial. For the Independent Commission on International Humanitarian Issues, the key is managing society; social equity will save the forest and human health will go with forest wellbeing. There will be modern technology, modernized health care and infrastructure. The disciplines invoked here are forestry, planning, engineering and medicine. The Brundtland report refers to "experts" on species loss and nature conservation, experts from the disciplines of biology and genetics. Prance starts from the viewpoint of forest production, how to modernize agriculture of the forests. All these views are not opposed; rather they are interlocking and the strategies proposed are complementary.

A common theme is the need to utilize the knowledge and resources of local indigenous peoples and not just rely on external expertise. This has implications for the practice of local democracy (Barbosa 1993) and, often, for the redistribution of property rights (Vandermeer 1991) over areas of significance for diversity; such changes are seen as preconditions for preserving those areas. Mott & Bridgewater refer to "the Aboriginal people, who understood and managed the land and its biodiversity for at least 60,000 years" (1992: 284) and Prance considers it is instructive to look at the way "Indians use forest diversity" (1990: 896).

So, there are calls for the greater participation of local peoples in decision-making for conservation and development (Del Amo & Ramos 1993, Wells & Brandon 1993) and for them to share in the benefits from new management practices (Martin 1992). In this way, social justice becomes a central concern of the new practice discourse of biodiversity (Cummings 1990). Social justice and political considerations mesh almost seamlessly with the conceptual and informational aspects of measuring biodiversity. This is a highly integrated discourse with many facets. It is perhaps the holistic discourse within the environet.

Differences and realignments

We have already referred to the problems of definition of the concepts of biodiversity and species, to the contested issue of measurement and the ubiquity of estimates of species loss.

> The term "biodiversity" applies to the variety and variability of living organisms from genes to elephants . . . The total number of known species or organisms is about 1.4 million, of which about 250,000 are plants, 44,000 are vertebrates and 750,000 insects . . . Arroyo and his colleagues (1992) estimate that about 10 million different species probably exist . . . Various estimates

put the probable loss of global plant species over the next 30 years at 20–25 per cent with over 35 per cent in the tropics at risk. (Grant 1995: 66–7)

The question of the scale of biodiversity drives the fundamental issue of whether there is a problem of biodiversity loss and, if so, how great a problem. Formulating the issue in this way removes much of the urgency from the debate. This contrasts with population discussions where the tone of urgency is prior to the details of the issue of scale and informs contributions from all sides of the debate. There are two polar positions regarding biodiversity. On the one hand there is the denial of the need or possibility of tacking the problem. This may be because there is no serious loss occurring, that reassuringly "natural" processes are at work (a relatively uncommon position). Or it may be that these "natural" changes, although not reassuring, are inevitable. On the other hand there is the belief in the ability of environmental management to deal with the problem. These positions can be represented as one dimension of an issues table.

Table 2.1 Perspectives on biodiversity.

	Ameliorative	Elegaic
Global	International/managerial	Grand narrative/species loss
Local	Indigenous/organic	Narrative of cultural loss

The ameliorative position suggests an improvable and protectable world. Thus, Grant (1995: 69) finds hope in the UN Biodiversity Convention:

This important convention establishes the principle of sustained management or protection of biological reserves, seeks adequate financial measures, . . . Every nation has to prepare a national plan for conserving and sustaining biodiversity . . .

Lugo et al. (1993) argue that life is resilient, species multiply, that they develop against the grain, and that management helps. And Giddings & Garcia-Castro (1991) promote biotechnology, to nurture biodiversity, whereas Hammer et al. (1993) advocate monitoring and management.

The elegiac dimension, in contrast, generally mourns the lost worlds of the Earth.

If anything remotely resembling that population–economic growth scenario is played out, with an acceleration of habitat destruction, most of the world's biodiversity seems destined to disappear. (Ehrlich & Wilson 1991: 760)

Two other major habitat classes of particular importance for bio-diversity are wetlands and coral reefs . . . Loss of wetlands has not yet been quantified in most countries but it may be severe: for example, over 90 per cent of the wetlands in New Zealand have been lost since the arrival of European settlers. (Murray 1993: 73)

Coral reefs were one of Darwin's study areas and this again empha-sizes the historical continuities in the discussions of biodiversity. Anderson (1990) also catalogues losses, the lost species of the Brazilian Amazon, which is another Darwin locale.

The second axis is designated global–local and this introduces the issue of interests and ownership: who has legitimate property rights over biodiversity? Some accounts accentuate the global:

Biodiversity is the global composite of genes, species and eco-systems. (Grubb et al. 1993: 75)

Many conservation activities yield global benefits. If biodiversity is conserved in a tropical forest, for example, it yields a benefit to people in other countries, either because they simply want it to be there, or because it helps sustain basic biogeochemical cycles on which human survival depends. (Pearce & Moran 1994: 17)

Despite the overriding importance of tropical forests at the glo-bal scale, it is important not to lose sight of the need to safeguard biodiversity both at regional and local scales and away from the tropics (Barkham 1995: 94)

Some span the global and local, as does McNealy (1993) in context of an economic analysis. Others, such as Luteyn (1992), promote local uniqueness, in this case referring to "unique and extremely diverse" Andean ecology. Global is not necessarily the reverse of local. When Dixon et al. (1993) or Wells & Brandon (1993) advocate protected local areas, locality is the theme, but the context is global; the area is valuable only in the global context.

Politically the question is: how do local interests count against global interests? How do local, indigenous interests in areas of diversity relate to globe-spanning commercial interests in the uses of that diversity and international institutions of regulation? Is there any such interest as the common interest, globally (Whatmore 1995)? Is there scope for the globalizing strategies demanded by Ehrlich & Wilson (1991) and Ehrlich et al. (1993)? These are often contentious, heated issues:

> Top-down approaches are as unacceptable in environmental conservation as in development. . . . until those organisations establishing initiatives on the conservation of biological diversity treat indigenous peoples with respect, there will be little hope for either the conservation of biodiversity or the protection of cultural diversity . . . (Gray 1991: 73)

Box 2.7 explores this further in the context of the rainforests.

And what about non-human interests, the interests of the life forms themselves (Singer 1990, Devall & Sessions 1985, Johnson 1991)? Some defend an anthropocentric view (Hammer et al. 1993: 97):

> In this paper we argue that the most important anthropocentric value of *biodiversity* lies in the functioning of ecological services supported by the interaction of the organisms, populations and communities, performed through the marine environment providing goods and services to human society.

and again:

> The key problem is not the preservation of a particular set of species now threatened by extinction, but the conservation of the resilience of those ecosystems on which human activity depends and the ability of those systems to continue to provide valued ecological services. (op. cit.: 103)

BOX 2.7 Controversies

Theme: whose forests?   Topic: rainforests

The forest needs the indigenous peoples   "The "myth of the vast Amazonian emptiness" described by Chase-Smith (1982) is a common feature of forest development plans in all parts of the tropics . . . Low population density in tropical forest areas is often not so much a sign of emptiness but more a sign that indigenous peoples are living in balance_{NPD} with their environment." (Independent Commission . . . 1986: 29)

"Our" forest destroyed by outsiders
"Gardening in the tropics nowadays means
letting in light: they've brought in machines
that can lay waste hundreds of hectares
in one day . . .
The animals are gone too
By the time they've cut
The last tree in the jungle only our bones
will remain . . .
Before you came, it was dark in our garden,
that's true."
(Senior 1994: 93, "Seeing the light")

International markets will save the forest   "Today Christmas Island is riding the wave of a tourism boom, with two-thirds of its 54 square miles (140 sq. kilometres) set aside as a national park,_{NPD} including Rainforest jungles with a treetop canopy up to 100 feet (30 metres) above the ground, and most of the surrounding reefs. The island attracts both gamblers and eco-tourists._{NPD}" (Associated Press, 25 December 1993)

The forests belong to the world   "Neither have the tropical rainforests yet been declared a common resource. Conservationists advocate that the world's rainforests be given international status . . . An international commission should be established_{NPD} for their management." (de la Court 1990: 94)

Others present an alternative ethic:

> We can, in fact, act with moral consideration not only toward other humans and toward animals, but toward species, ecosystems, the biosphere, and toward anything else that may turn out to have interests. (Johnson 1991: 262)

> Only those who prefer religious faith to beliefs based on reasoning and evidence can still maintain that the human species is the special darling of the entire universe, that other animals were created to provide us with food, or that we have divine authority over them, and divine permission to kill them. (Singer 1990: 206–7)

> The natural preference for one's own species does exist. It is not, like race-prejudice, a product of culture. It is found in all human cultures, and in cases of real competition it tends to operate very strongly. We can still ask, however, how far it takes us. (Midgley 1992: 128)

So, the discourses of biodiversity move fairly effortlessly from conceptual definition, to information and measurement, to the consideration of global and local practices; but this is not to suggest that the discussion is in any way consensual. There are deep divisions in the discussion between recognizing or denying a problem, between the global and the local, between property owners and those without recognized rights, between human claims and those who speak for all species. Biodiversity may be a holistic topic but it remains essentially contested for all that.

Multiple interpretations

No single interpretation encompasses this first sequence: resources and energy, population, biodiversity, species, rainforest. The topics are multiple, they carry many meanings, they are *made* by many interpretations, competing interpretations. In a post-Malthusian view, the great linkage binds resources to population; more hopeful connections are also made, connections between human resourcefulness and the populated Earth. Others views connect biodiversity and population, two aspects of the measurable Earth. Quantities rise and fall: resources, human numbers, species, rainforest areas. Evolution moves forwards; thermodynamics gradually fades out. Modernity is a threat to nature; modernity is a hope of a better future, based on science or on improved technology. The ownership of the Earth is at issue: local or common; national or multinational or international. Rights compete; national rights, indigenous rights, human rights, rights of women, rights over knowledge, species rights. The sequence is about many things: hope, modernity, quantity and value; the unity and diversity of science, rights and interests across the Earth. Each reader finds a different nuance: that is why environment is so richly discussible.

CHAPTER THREE

Topic analysis (II)

Pollution, polluter

Cultural innovation

We considered the inputs into the economy–environment cycle in the form of resources at the start of Chapter 2; we now consider the outputs, the issue of how waste products are disposed of and the negative impact termed pollution. If the environment provides us with goods in the form of energy, minerals and natural features, then it also provides services in the form of pollution sinks, whether into land, water or the atmosphere. Here the ecological cycles provide a mechanism by which waste products are stored, dispersed or reconstituted. The issue is when the results of these processes still leave a negative impact in terms of the effect on human health and natural environments.

Because we constantly generate waste products, and the environment is constantly implicated in their disposal, the possibility of pollution is also ever present. The question of identifying when pollution is occurring invokes new information discourse, and a new information discourse of a specific kind: the new information of disclosure. An illustration of pollution's new information discourse of disclosure is provided by the media coverage of the Exxon Valdez oil spillage in the Prince William Sound, Alaska, as it happened in 1989.

Box 3.1 shows both the emphasis on providing new information, often anecdotal but immediate, on the oil spillage and also the variety of concerns that the news item is linked into: ethical, technocratic, economic. In this way the term "pollution" is put into different contexts as the pollution event is disclosed; that is, there are different

BOX 3.1 New information discourse

Theme: disclosures of the Exxon Valdez disaster
Topic: media bite on pollution

Ethical/naturalist agenda "The 11-million gallon spillage, one of the world's worst oil pollution disasters, is decimating marine life off the Alaskan coast." (*Today*, 27 March 1989)

Ethical/naturalist agenda ". . . The rare humpback and killer whales that inhabit the area are also expected to die from the pollution of their food supplies . . ." (*Sunday Times*, 2 April 1989)

Biodiversity agenda: species loss "Environmentalists were yesterday battling to save the beautiful beaches and calm waters of Prince William sound, home to many endangered species. At risk are ten million migratory birds . . ." (*Today*, 27 March 1989)

Amenity agenda: wilderness loss "The exploration is already being fiercely contested by the environmentalist lobby which has strong support among those choosing to make their homes in the Alaskan wilderness." (*Sunday Times*, 2 April 1989)

Economic agenda "The oil spill that still spreads its contamination across Prince William Sound in south central Alaska has also inflicted severe damage on the [oil] industry at a time when it faces a crucial decision for the future." (*Guardian*, 3 April 1989)

Social–cultural agenda "While scientists ponder the distant consequences of the Exxon Valdez's Easter mishap, and Exxon ponders how best to staunch the flow from the hole in its haemorrhaging reputation, Dr Thomas Nighswander has a different concern. As a doctor with the Alaskan Area Native Health Service, he is monitoring the effect of the oil-spill on the already fragile economies and culture of the four isolated fishing villages, whose total population is fewer than 1,000 people." (*The Economist*, 26 August 1989)

Technocratic agenda "The 10m gallon oil spillage from the Exxon Valdez tanker in Alaska has focused attention on the methods used by oil companies to deal with such disasters. Although efforts to stem the pollution from the Valdez appear largely ineffectual, work is going on to develop a better procedure for such emergencies . . ." (*Financial Times*, 5 April 1989) agendas within the disclosure itself. The disclosure is already contested.

The ethical concerns reminds us that pollution also has a spiritual sense: the defiling of Earth, the loss of origins. Here there is an ancient mythical discourse, the inverse of a new concept discourse. The debates reverberate with primary conceptions of radical myths, conceptions of clean and dirty, pure and stained. This reminds us of the cultural dimension to the perception of pollution issues (Douglas 1966, Vincentnathan 1993). It also links into the types of issue associated with an ethical definition of the problem of pollution.

"Pollution" invokes new information discourses in so far as pollution is about precise claims and measurements, although one might query the extent to which it is *new* information that is provided. In many cases what is offered is the development of more precise forms of previously known information or the application of existing knowledge to new materials, processes and locations. And this is not the construction of new knowledge but rather the disclosure of previously held knowledge as specific elements of information.

Hence, this strong element of disclosure provides many opportunities for polemical discourses around new information.

On 23 October 1989, Paul Brown of *The Guardian* newspaper reported that the inhabitants of the Yorkshire village of Drax were suffering from a strange new form of pollution. Sewage bacteria were raining down on their homes in a fine dust. A community medicine expert from York warned that the silt might contaminate food. The immediate source of the pollution was steam from the giant cooling towers of the Central Electricity Generating Board's Drax power station, which took water from the River Ouse. The power station had fallen victim to one of Britain's sewage-laden rivers. Untreated sewage was being poured into the river upstream, with more swilling up from the Trent on the tide. (Rose 1990: 1)

Discussions of pollution are not simply about demonstrating that harm has occurred. They lead forwards into demands on public policy. Pollution is on the agenda for official decision-making by central and local governments and by international institutions such as the UN. This invites a new practice discourse.

Armed with policy instruments suitable for the Victorian era of choking, dust-ridden air within industrial towns, and rivers so grossly polluted that people ran from the banks, Britain triumphed over the London smogs and cleaned up the Thames, though not until the 1960s. Then in the 1970s and 1980s, the United States and Europe developed new policies and methods to match a whole new range of more insidious and more complex environmental problems. Britain, however, sank into the deep sleep of the just, or at least the self-satisfied. (Rose 1990: 4)

Politics of Pollution – A bill pushed through the Senate's judiciary committee last week would require the authorities to justify any significant regulation of business with complex risk-assessment studies and cost–benefit analyses. (*Financial Times*, 3 May 1995)

So, in the discourses around pollution we find discussion of novel methods for dealing with the problem, both technological and economic. In terms of the technology of new practices, there is significant concern with abatement technology and with monitoring. These are both marginal adjustments to the processes that generate the pollution as compared, say, to consideration of alternative production methods, or, more radically, alternative production modes.

They are organized around several key terms: best practicable means (BPM) and its new invocation best practicable environmental option (BPEO), the cornerstone of integrated pollution control (IPC):

A BPEO is the outcome of a systematic consultative and decision-making procedure which emphasises the protection and conservation of the environment across land, air and water. The BPEO procedure establishes, for a given set of objectives, the option that provides the most benefit or least damage to the environment as a whole, at acceptable cost, in the long term as well as in the short term. (RCEP 1988: 5)

and also "best available technology not entailing excessive cost" (BATNEEC) (RCEP 1988: 6). In all these new practice proposals the position of the regulator is paramount.

In terms of the economic approach to new pollution control practices, there is now an extended discussion of the merits of market-based instruments (Markandya & Richardson 1992: Part II; and also examples such as Stavins & Whitehead 1992, Tahmassebi 1992). This is framed in terms of argument and counter-argument between regulation and the new instruments.

The disagreement takes the form of considering the relative advantages of regulation or command-and-control as against market-based instruments, taxes, permits, subsidies. The market-based approach can be couched in technical economic language with the "proof" that social welfare is maximized at least cost by taxes or permits rather than any other means. This works by demonstrating that taxes provide a cost-effective incentive to innovate and introduce technology abatement equipment; one outcome is a shift from control to abatement as the key focus.

The practice of deriving policy measures from environmental economics also encourages other new practices. Pollution control requires measurement, and it requires economic measurement as well as scientific measurement. The implicit issue is how to connect quantity, in terms of amounts of pollution, to quality, the quality of life. We therefore need new ways to measure pollution, ways that connect science and economics. The result is a call for the new practices of accounting and valuation (Markandya & Richardson op. cit.)

It might be thought that this terminology in technological and economic areas – BPEO, BATNEEC, MBI (market-based instrument) – offers a new concept discourse. But this is not really the case for two reasons. First, most of this terminology is not really new or even a reinvention but rather a straightforward re-use of old terms; BPEO develops from BPM, which has been in use since 1874 (RCEP 1988: 1); the economic terminology has long roots in neoclassical welfare economics. Secondly, a new concept discourse involves a stimulus to a completely new way of thinking about an issue or problem. The kind of terminology offered in pollution discussions is not of this kind. In most cases it involves a continuation of the technocratic approach to pollution, perhaps redefining the technocrat to include economists as well as environmental scientists, but without proposing radically alternative views of generating, managing and reducing pollution (Commoner 1991: 101–2).

Topic analysis (II)

Differences and realignments

"Pollution" is nevertheless a polarizing issue associated with antagonisms, antitheses, and conflicts. When people use the word "pollution", they expect others to disagree with their views. These disagreements on the level of argument relate closely to the conflicts of interests that the polluting activity itself generates. For pollution is classically defined as an externality, where the activities of one impinge on the activities of another without any exchange of payment or compensation (to adopt a simplified definition). Therefore, pollution must create two parties with a conflict of interest: the polluter (actual or potential) and the polluted.

These two positions can be associated with two opposing points of view. The polluter can claim that pollution is a necessary by-product of beneficial economic development and that this development will, in due course, generate the resources for cleaning up the pollution. In this view, the removal of regulatory or market-adjusting mechanisms is advocated in the belief that entrepreneurial incentives will encourage the limitation of pollution through consumer demand and business ethics. This emphasizes the market potential of green products and technology and rejects the notion of widespread market failure. Thus, in relation to the debate about the North America Free Trade Agreement, the Mexican Commerce Secretary, Jaime Serra Puche, is reported as having said:

> ". . . NAFTA will help to raise pollution standards and bolster the cleanup effort . . ." "Mexico Asks: Will Pollution, Drugs Roll in with Duty-Free Goods?" (Associated Press, 28 December 1993)

The position of the polluted, by contrast, represents the view of pollution, any pollution, as ethically wrong. This ethical position could also be categorized as that of non-scientific common sense. This would see any form of emission as pollution and therefore see pollution control as being about preventing any such emissions. The emphasis is on the moral responsibility to avoid pollution and to clean it up. At one level this responsibility can just be about dealing with the dirt one creates; at another level it is linked into a greater responsibility of human society in relation to the natural environment. Within the ethical positions, there are humanists and naturalists; humanists emphasize social dimensions, naturalists the wilderness aspects.

Certainly, this ethical standpoint provides the tightest use of the precautionary principle in terms of avoiding damage from pollution where any doubt exists (RCEP 1988: 12 and Appendix 3).

But the map of viewpoints and associated positions is more complex than a simple dichotomy, the polluter versus the polluted. First, it is not always clear who should be allocated to which category, as when environmental economists seek to argue that consuming a good, such as electricity, which has caused pollution in its production, turns consumers into polluters.

> Making the consumer of the polluting product pay some of the clean-up cost may seem at odds with the PPP but in fact it is exactly what should happen. (Pearce et al. 1989: 158)

And sometimes pollution is so generalized in its impact that the polluters come to see themselves as part of the polluted group. This raises general questions of lifestyle.

Secondly, there are other categories of groups who would benefit from greater regulation or control of pollution. These include the regulators themselves, institutionally based within professions, government or quasi-governmental agencies. And there are sectors of industry that do not pollute, or pollute on a lesser scale than their competitors, and would welcome such "unfair competition" being removed by the level playing-field of environmental regulation. Another industrial sector that would benefit from pollution regulation is the abatement technology sector. These more complex crosscutting positions in relation to pollution provide support for two other viewpoints: technocratic and economic.

The technocratic positions are associated with the scientific analysis of pollution and the technological management of that problem. The environmental scientist position would talk about critical capacity, thresholds and absorptive capacity, and would consider that pollution only occurs once the level of emissions is causing damage in a particular environmental system (e.g. Philipp & Bates 1992, Harrington et al. 1993, Jarallah et al. 1993); this rather begs the question of what "damage" is. Hence, there is an emphasis on epidemiological studies of the impact of pollution and associated discussion of the statistical validity of such inferences. The term "toxicity" is frequently used in this context to describe materials with polluting consequences (e.g. Twitchell 1991, Renzoni 1992, Easterley et al. 1993). These approaches are particularly prevalent in discussions of water pollution.

The presentation of scientific material and the role of scientists themselves are central issues here. Scientists may measure, experiment and observe. But ironically, "pollution" produces *arguments* because science is at stake. When people claim "pollution", they introduce science into the media and the public discussion. This sounds like a cue for agreement, but scientists often disagree, and their differences can be more intense than ordinary differences since their authority is at stake. And whose science is speaking, on any particular occasion? Science was considered neutral and objective, but now people suspect motives, they ask questions about who controls science (Taylor & Buttel 1992, Hart & Victor 1993, Wynne 1994).

On the other hand, one finds the emerging dominance of economic approaches to defining pollution and a close link into specific policy proposals. The environmental economist would answer the question of "What is damage?" using the concept of optimal pollution; this is where the marginal benefits (usually measured in terms of profit or wealth generated) of polluting just balance the marginal costs of the environmental damage caused (Pearce & Turner 1990, Markandya & Richardson 1992). Using an optimal pollution concept raises not only the conceptual issue of whether this is an adequate valuation base but also the technical one of valuation methods. Contrasting with water pollution, air pollution seems particularly likely to invoke this economic approach.

Moving beyond the level of defining pollution, the economic approach to pollution has placed the terminology of market failure and of market-based instruments to redress that failure at the heart of the policy discussion. Although the precautionary principle may be invoked to garner support for policy measures, the core principle used here is the "polluter pays" principle, since this involves the question of how much the polluter must pay, what is the value of pollution both in terms of the economic output and welfare generated, and the valuation of damage to the environment:

The principle to be used for allocating costs of pollution prevention and control measures to encourage rational use of scarce environmental resources and to avoid distortion in international trade and investment is the so-called "Polluter Pays Principle". This principle means that the environment is in an acceptable state. In other words, the cost of these measures should be reflected in the cost of goods and services which cause pollution in production and/or consumption. (Pearce et al. 1989: 157 [interpretation of OECD 1975]).

Pollution polarizes viewpoints along many axes, interconnecting with disputes in modern culture, disputes about production, knowledge and expertise, modernity and lifestyle. The polarization is into multiple disputes. There is no unified dispute, there are many different disputes, and they don't polarize symmetrically, with predictable sides. The pollution map is one of many self-contained disagreements, differences, going on in parallel but with only limited connections between them. At the theoretical level, there are different discussions, each with its own polarization: for example, the technological discussion of pollution and the economic argument about MBIs. And these internal debates are gaining prominence: for example, command and control versus taxation; between methods of applying IPC, BPEO and BATNEEC; and between valuations of environmental "damage".

The different discussions borrow from each other, or rather the economic analysis borrows from the technological, but only selectively. Adding policy and media debates generates more polarizations, connected but not unified. These are localized polarities, not amounting to ideological fronts across the whole pollution agenda. The polarities are defined locally, at each node on the network of issues.

Global warming, climate change, the greenhouse effect

Cultural innovation

Global warming due to the buildup of greenhouse gases is not an isolated problem. It is currently the most prominent, and over the long term probably the most significant, of a new class of environmental problems. (Mintzer 1992: 9)

The kind of movement that saved the Grand Canyon – is it

capable of addressing global problems like carbon-dioxide emissions?

I think it is. The whole thing – the excitement over global warming, acid rain, hole in the ozone, loss of species – is getting through the way it never did before. (David Brouwer quoted in McKibben 1992: 243)

Global warming and related issues act as a discursive litmus test. They have come to stand for the new environmental agenda and the advancement of the environment in global consciousness. Global warming is therefore a symbolic issue, as well as a practical one. It symbolizes the changed world, and even sceptics are affected by the idea of the world being different:

The attractions of a radical perspective are very real . . . The signals of impending disaster are many. Global warming and especially the depletion of stratospheric ozone are extraordinary threats to life on earth . . . (Lewis 1994: 5)

If there is to be a scene-shifting impact on politics and culture, then global warming will be a lever of change. Can the agendas be reshaped and if so how? What discursive resources can this process of change draw on?

Although the "greenhouse effect" or "global warming" may have been new terms for many at the time that the scientific literature and the work of the Intergovernmental Panel on Climate Change was first making its new findings known in the late 1980s (IPCC 1990), it cannot be said that the discourse around these terms now has the character of a new concept discourse. In fact the point is frequently made in the literature that global warming and the greenhouse effect are "old" processes. This is not to deny that the "global warming" debate is about newness, a strange new world. However, this is a world gradually growing strange. There is not the imminence of the population arguments, notably, where trends are felt as more immediately present. The same is true for some pollution arguments and, above all, for biodiversity arguments, with their emphasis on species loss now, rainforest loss now. Global warming is a problem poised on the horizon, indicating a separation between the present and a future world that *will* be different.

The handling of time is therefore very important in global warming arguments, and new information discourses are highly significant in these issues of change over time. New information about the greenhouse effect is central here; indeed, the recognition of the phenomenon is based on new information:

> The advanced track scanning radiometer will monitor sea surface temperature and land vegetation, to help scientists in the early detection of global climate change into the next century. (*Herald,* 22 December 1993)

The status of the information itself is a key aspect:

> Sceptics holding out against the idea that the world is warming will have a harder time dismissing the latest predictions from British meteorologists. A team at the Hadley Centre for climate prediction and research, based at the Meteorological Office in Bracknell, has developed a computerised climate model that has accurately reproduced past fluctuations in the Earth's temperature – the first time a model has succeeded in doing this. This breakthrough counters the sceptics' arguments that computer models cannot be trusted. It also confirms fears that the world is warming at an unprecedented rate. (*New Scientist* 26 January 1994: 6)

Such a discourse is necessarily scientific, the texts must refer to science or technology, they must claim such expertise, or they must cite other experts. In the passage from the *Herald* above there is a celebration of advanced technology and new techniques of measurement.

The new information discourses also have a strong fact-disseminating character, an expert talking to others, or an expert being cited by others. The global warming debate is a place in culture where the voices are authoritative: they announce the data, they declare the implications, as in the announcement of a public talk below:

> The WMO Commission for Climatology, as one might guess, is concerned with monitoring the global climate. Its president is Dr John Maunder ... Dr Maunder calls his lecture "Policy aspects of climate and of climate change" ... He will present

evidence of the potential value to a country of detailed statistics on its climate; he will provide examples of ways in which such information can be used effectively, at Government and at business levels, as an important input into decision-making; and he will also argue that any change in climate that might come about as a result of global warming, can, if managed properly, be transformed into a resource to improve the economic performance of a nation. (*Irish Times*, 10 December 1993)

At the heart of the global warming issue is the recognition that the issue only exists because of the scientific measurements and because of the ability of the new techniques to deal with lengthened time horizons. In global warming arguments, science is remaking the cultural significance of time itself (Adam 1994).

Uncertainty is integral to such knowledge of future long-term horizons. There is uncertainty about both the processes of change and the scale of the impacts:

> **Figures Point To Global Warming: Temperature Measurements Show This Year To Be The Sixth Hottest On Record**
> The world was a warmer place this year compared with 1992, according to figures released yesterday by scientists investigating the greenhouse effect . . . "The nub of the problem is that we cannot yet prove that it's caused by man-made greenhouse gases," Dr Bennetts said. (*Daily Telegraph*, 22 December 1993)

> Uncertainty refers to a lack of knowledge that further information can provide. Currently, significant gaps exist in our understanding of the climate system and how increasing concentrations of greenhouse gases (GHG) may affect it. Similarly, our understanding of natural and socio-economic systems that both determine and bear the effects of global climate change is incomplete. (Rothman & Chapman 1993: 89)

Many of the more popular or policy-orientated discussions start from the premise of major physical impact and social significance (see below), but the central place of information discourses from the scientific literature in these discussions means that the question of how certain the impacts are cannot be ignored. This undermines many of the more normative and polemical arguments that are made for policy intervention and can generate a tension between justifying intervention and recognizing the scientific material that is the original justification for global warming being on the policy agenda at all.

But on the policy agenda it certainly is, and, as a result, new practice discourse becomes a key resource. The form of this new practice discourse is not technical to match the scientific justification for policy; there is relatively little consideration of the technological aspects of achieving reductions in the identified greenhouse gases (and this contrasts with the situation on CFCs and the implementation of the Montreal Protocol). Rather, the emphasis is on the problems of achieving policy agreement in the first place. Hence, the new practices involved are practices of negotiation and diplomacy. Games, zero-sum or otherwise, are often invoked to explain the process of achieving such agreement and there has been a recent use of game theory or similar frameworks to understand the positions of participants (Black et al. 1993, Martin et al. 1993, Ward 1993). The contradiction between individual pain and collective benefit is the central theme, with the analyses using the models of games such as prisoner's dilemma or chicken, of the collective action problem and of public goods.

This introduces a concern with institutional rationality, relating to how organizational actors interact, and this will be different from the scientific rationality invoked in new information discourses (Hart & Victor 1993). A tension results, possibly even a contradiction:

> So far, governments seem unable to agree a policy response to global warming that matches the undoubted global threat spelt out by IPCC's scientists. Yet the IPCC was an unprecedented process of international scientific collaboration. (Leggett 1992: 38)

The global warming discussions are therefore significant culturally, because they connect reliance on scientific expertise with these new institutional rationalities, and because they raise and develop issues of uncertainty and long-term time horizons, addressed both through technological authority and political strategy. Global warming draws together two arguable worlds in contemporary cultural thought, the arguable worlds of science and environmental diplomacy. The result is friction, since science cannot resolve the disputes; indeed science tends to fuel the arguments on different sides. Yet political negotiation cannot do without the science; a need remains for scientific legitimacy to justify states' actions. The arguments from science and diplomacy, the new information and new practice discourses, therefore circle each other, like the paths of two orbits around a planet, crossing at times, then passing out of sight of each other. There is no resolution of such arguments.

Differences and realignments

These discourses are used to argue two key questions concerning global warming: the enhancement of the greenhouse effect and potential climate change. What exactly is happening? And who will be affected?

The underlying processes of global warming – the build-up of various gases such as CO_2, methane and CFCs within the Earth's atmosphere, leading to enhanced insulation and a higher proportion of the Sun's energy being reflected back to the Earth's surface – are widely regarded as well established in the popular, policy and academic literature; but, as indicated in the introduction of new information discourses, there is much more doubt about the scale of resulting changes in average temperature levels and sea-level rise, for example. There is also doubt as to the significance of the physical changes for society.

One key focus is on the issue of scale. There are four discernible positions (Table 3.1).

Table 3.1   Perspectives on climate change.

	Low significance	High significance
Moderate change	Sceptical	Precautionary
Substantial change	Adaptive	Crisis-warning

The press, the policy debate and the social science academic discussion almost all start from the premise that we are in the bottom right-hand corner of this table, that there is something major to discuss:

> Today, the world is faced with risks of irreversible damage to the human environment that threaten the very life-support systems of the Earth (Gayoom, President of the Republic of the Maldives 1991: 14)

The magnitude and extent of the changes that could occur defy our imagination. (Lalonde 1991: 82).

From this point, policy can be devised, discussed and critiqued. There is more measuring of policy against general sustainability criteria *on the assumption* that the enhanced greenhouse effect is undermining sustainability (crisis-warning), than testing of the policy proposals against specific predicted climate change impacts (adaptive, precautionary), let alone questioning the need for any such policy (sceptical).

A second focus is on distributional issues: who will be most affected by the physical changes; who is responsible for causing those changes; who should bear the costs either of the physical changes themselves or of policy measures to avert and mitigate those effects? In part this is an issue of the extent to which private property owners (or their insurers) rather than the state should bear these costs; in part it is a restatement of the "polluter pays" principle to suggest that all polluters, whether producers or consumers, should bear the costs – with a counter-argument that the burden of these costs could be excessive or regressive in their effects. In debating these questions, there is a clear role for new practice discourses of the kind outlined above.

There is a strong North–South dimension to the discussion (Hyder 1992). Since developed countries have caused the problem through past industrialization, should they not bear the burden by cutting back on their growth so that developing countries can catch up (Mukherjee 1992, Lave & Vickland 1989)? "We did not contribute to the impending catastrophe to our nation; and alone, we cannot save ourselves." (Gayoom 1991: 16)

Or should some recognition be given to the status quo in development so that future rights to emit these gases are in some sense related to past practice?

Thus, the Climate Change Convention itself indicates areas in which further negotiations are urgently required. . . The specific commitments of the developed countries, including the East European industrialized countries . . . Without doubt, these countries have appropriated more than their due share of the planet's atmospheric resources and have thereby induced the

phenomenon of incremental global warming with all its attend-
ant dangers for humankind. (Dasgupta 1994: 145–6)

As by far the greatest increase in greenhouse gases is caused by
the combustion of fossil fuels, mostly in the rich countries, the
inundation of low-lying cities in the developed world would thus
be a self-inflicted disaster. But many cities in the developing
world also have coastal locations sensitive to flooding. (Girardet
1992: 68)

So, in global warming discussions we have tensions between scientific expertise and the rationality of policy negotiation; we have difficult questions concerning uncertain impacts over a long timescale
and the distribution of costs and benefits.

This takes us to why global warming arguments are culturally central. They pose questions of expertise, and behind that is a crisis of
rationality. The different experts represent different rationalities, in
each of which facts mean something different. There is a multiplication of expertise, a diversification of rationality, and issues of disciplinarity are raised, as illustrated in Box 3.2.

There is a diversification of claims to expertise. This diversification can be read as competition, between hierarchy and opposing
hierarchy. It links to the question about the hierarchy of rationality
itself. Is natural science at the summit of rationality? Or social science? Or economic science? Can they integrate or do they oppose
each other? Such questions are not answered by the global warming
discussions, rather the discussions are framed by such questions.

This results in an unusually self-reflexive discussion in which the
impact of global warming on politics and culture is central. Box 3.3
illustrates some of this discussion. Here the controversy is reflexive:
it is about the arguments, their impacts, and the discursive process
itself. For Gore, the arguments advance slowly, the proponents
struggle, and campaigns mount, whereas for Dasgupta, the institutions have already responded with the international agencies redefining the agenda. For Bramwell, there are two impacts. One is
reasonable, with the spread of information and people responding,
when it is relevant. But on the other hand "ecological movements"
exaggerate, they are making capital, and so the cultural impact is
clouded by political self-interest. For Gore, the environmental

BOX 3.2   Controversies

Theme: divergent expertise   Topic: global warming

Cloud science versus computer simulations   "Clouds might help even out the effect of global warming worldwide, countering an exaggerated warming towards the poles that has been <u>predicted by computer simulations</u>,_{NID} a study says." (Associated Press, 15 December 1993)

Climatology versus economics   "William Nordhaus, professor of Economics at Yale University, was the first to try to supplant scientific concern with economic scepticism . . . It is cheaper to adapt to climate change than it is to stop causing it. Their views, predictably, have been adopted by the fossil fuel industry . . ." (Wysham 1994: 205)

Social sciences versus physical sciences   "This dominance of physical climate research over institutional analysis points to the second issue, <u>the hierarchy of the physical over the life and social sciences</u>._{NPD} This hierarchy constitutes an environmental determinism: the physics and chemistry of climate change set the parameters for environmental and biological change . . ." (Taylor & Buttel 1992: 410)

The divided communities of science   "In 1992, Greenpeace conducted an opinion survey of the international climate science community. It suggested that in a world which makes <u>no efforts to cut emissions of greenhouse gases</u>,_{NPD} around half of the scientists rank the <u>worst-case analysis of unstoppable heating</u>_{NID} as a possibility. Furthermore, a depressingly large minority – more than one in ten – hold that a thermal runaway would be a probability." (Leggett 1993: 46)

Natural scientific and social scientific models   "In the example of global warming, many scientists around the world have been engaged in the <u>measurement of global temperatures, CO_2 emissions, rising sea levels etc.</u>,_{NID} to find out if the greenhouse effect is really observable. But their findings have not been directly related to possible consequences of the greenhouse effect on society and future generations." (Yamaguchi 1993: 78)

BOX 3.3   Controversies

Theme: degrees of cultural and political impact
Topic: global warming

The long struggle for impact   "Over the next few years, I began seriously studying global warming and several other difficult environmental issues . . . But
despite mounting evidence that the problem was truly global, <u>few people were
willing to think about</u>$_{NCD}$ the comprehensive nature of the response needed . . .
Most people still thought of the environment in local or regional terms, so it
was impossible to get adequate <u>funding for research</u>$_{NID}$ on global warming."
(Gore 1993: 6–7)

The sudden impact   "Few issues have acquired <u>priority in the international
agenda</u>$_{NPD}$ in as short a time as global warming." (Dasgupta 1994: 129)

The rational impact of information   "Mounting concern about human interfer-
ence with the Earth's atmospheric heat balance, and associated predictions of
climatic change, led to <u>the establishment first of an intergovernmental scientific
panel and then to negotiations</u>$_{NPD}$ on a Framework Convention on Climate
Change . . ." (Grubb et al. 1993: 8)

The double impact: real and exaggerated   ". . . and the greenhouse effect,
while still in the realms of prediction, has <u>galvanized governments into action</u>$_{NPD}$
. . . If there had never been an ecological movement, and evidence had
emerged to suggest that the greenhouse effect existed, and that some countries
might be deleteriously affected by a greenhouse effect, then those communities
who thought they would be affected by it in a bad way would be lobbying . . .
There would be protests and there would be action. The protests would not
be tinged with righteous hysteria, as now, and we would be much clearer as
to what exactly was happening . . ." (Bramwell 1994: 167)

The muffled impact: failures of cultural communication   "The scientists of the
IPCC have undersold the <u>worst possibilities outlined</u>$_{NID}$ in their analysis of global
warming. And the policy-makers have misunderstood fundamentals of scientific
method." (Leggett 1992: 38)

96

agenda advances because of science, new information and rational political argument. For Dasgupta, there is science and then there are rational administrative institutions. For Bramwell, there is rational politics, and then there is irrational politics.

So the arguments around global warming are characterized by conflict, conflicts of expertise and of rationality; conflicts of appropriate political response and future policy direction. And a central element in this is the status of scientific expertise, the status of policy institutions and the relative status of interests distributed across the warming globe.

Sustainable development, sustainability

Cultural innovation

In June 1992, the United Nations' Earth Summit in Rio pushed the concept of "sustainable development" to the top of the agenda, adding to public pressure on governments to be seen to be green. (*European*, 17 December 1993).

Mrs Brundtland said . . . Unless the world moved towards sustainable development, the east–west confrontation which had gone would be replaced by a north-south divide. (*Guardian*, 25 June 1992)

Ten years ago, it would have seemed improbable that leading international business people had even heard the term "sustainable development". In ten years' time, current business attitudes on this crucial issue will perhaps seem improbably timid and naive. (*Guardian*, 8 June 1992)

These media responses to the 1992 Rio Earth Summit (or United Nations Conference on Environment and Development) typify the presence of "sustainable development" in public responses to environmental issues. The emphasis is on the new: new consciousness, new crisis, new remedies. The implication is clearly that there is a new environmental agenda: people perceive new environmental problems and they also propose new answers.

It is hoped that future aid projects in developing countries will be drawn up in consultation with local people so they promote sustainable development rather than merely economic growth. (*European*, 31 December 1993)

A point on which everybody agrees is the need for "sustainable development". In principle, this idea is clear enough: development, not merely economic growth should be the target, and it should lead to enduring improvements in welfare. (*The Economist*, 25 December 1993)

The terms "sustainable" and "sustainability", often used interchangeably, have a central place in this agenda. They impact as new concepts, supported by new concept discourses. As new concepts then arise both to define the ills and to articulate the answers of society. Indeed, the discussion requires conceptual innovation. Since the agenda is new, dated language is not convincing, it sounds irrelevant. Sustainability and sustainable development often present themselves as the most influential concepts in the environmental agenda because they stand for new thinking, and they have been echoed and re-echoed in many contexts, specialist and political, scientific and economic, global and local. But "newness" is not a simple idea: what is new in one context is familiar elsewhere; and a successful concept recurs so often that it ages quickly and has to be actively renewed.

Key passages show the nature of this new concept discourse. The Brundtland Report provides the focal definition to launch the new concept discourse:

Humanity has the ability to make development sustainable – to ensure that it meets the needs of the present without compromising the ability of future generations to meet their own needs. (WCED 1987: 8)

This definition is taken up in the UK government's *Strategy for sustainable development,* for example:

The question however is, do we have to wait until disaster overwhelms us before we make the radical changes necessary to protect our world for future generations? That is the vital challenge of sustainable development. (UK Government 1994: 5)

itself echoing work by David Pearce and his team:

The term "sustainable development" itself ought not to occasion much controversy. (Pearce et al. 1989: 1)

While it is a popular pastime to collect different and incompatible definitions of sustainable development, inspection of the words and of their origins suggests that defining sustainable development is really not a difficult issue . . . What is being referred to is sustainable *economic* development. (Pearce 1993: 7)

Sustainable development shifts the focus from economic growth as narrowly construed in traditional attitudes to economic policy. (Pearce et al. 1989: 21)

Here we find the emphasis on the future, the definition of the relation between economic or development processes and environmental processes, and – in the Brundtland extract – the implication for equity concerns.

In this conceptual context the very ambiguity of "sustainable development", which is often remarked on (Hart 1994: 705; Murdoch & Clark 1994: 115), becomes part of the process of enriching and renewing the concept, and not simply a problem of implementation (Grove-White 1994: 29). It is true that substantial argumentative effort within the Pearce *Blueprints* (Pearce 1989, 1993, 1995) and similar texts (Turner 1993b) is focused on developing policy tools in pursuit of sustainable development goals. However, most of the emphasis is on reconceptualizing, on devising new ways of thinking (although largely from within established disciplinary bases) and the details of new practice are often left to be worked out in more precise contexts, such as pollution control. The sustainable development discussion remains first and foremost a new concept discourse aimed at altering ways of thinking.

This is illustrated in Box 3.4, which brings together examples from the newspaper media. Everywhere, it appears, sustainable development promotes new thinking, or the idea that new thinking is required. For this reason, sustainable development is important to new concept discourse generally, that is, in the argumentative culture of the time. Even where the response sounds sceptical, as in the report of the "new religion" of sustainable development, the impact persists: something has changed, however we assess it; new beliefs and values are spreading.

BOX 3.4   New concept discourse

Theme: the 1990s as new era
Topic: media bite on sustainable development

New politics   "In addition to shifting its focus from holding the line against communism to promoting "sustainable development" abroad, the foreign assistance agency has faced persistent charges of waste . . ." (*Washington Post*, 20 November 1993)

New ethics   "Top BSR [Business For Social Responsibility] priorities include: finding a way to create a more human workplace, increasing awareness of environmental impact, pushing for more direct community involvement, and replacing the idea of constant expansion with the environmentally kinder theory of sustainable development." (*Washington Post*, 22 November 1993)

New economics   "In 1989 the World Bank began to screen projects routinely for environmental effects. Since then it has become greener in other ways . . . It has created a new vice-presidency of environmentally sustainable development . . ." (*The Economist*, 25 December 1993)

New religion   "Aracruz [eucalyptus pulp company] is a leading exponent of 'sustainable development', the new religion enshrined at the 1992 Earth Summit." (*Guardian*, 10 December 1993)

Sustainable development represents new ethics, new politics, and new economics. Of course, the question is: What will the newness amount to? Is it a genuine offer of change or mere sloganizing? So, the issue of delivery arises: What will the concepts actually achieve? New concept discourse invites scepticism, as well as enthusiasm and creativity.

One area where new practice discourses merge with the new concept discourse is in the attempt to formalize the concepts of sustainability and sustainable development in indicators.

Indicators have to be developed in order to measure how sustainable a nation's economy is. Sustainability indicators are not intended to be snapshots of the environmental situation, but to

be normative . . . The statistics "indicate" sustainability levels.
(Bramwell 1994: 156–7)

Here the assumption is that either the process of defining sustainable
development has been completed or that it can no longer proceed at
the conceptual level but must be achieved through specifying it, pref-
erably quantitatively. Hence, work by Pearce (1993) and the New
Economics Foundation (MacGillivray & Zadek 1995) among others
seeks to give the concept fairly immediate policy relevance by creating a usable tool. This parallels the use of such indicators in other
areas, notably social policy and financial practice (Horn 1993,
Massiah 1993, Tainer 1993, World Bank 1993).

But the emphasis on indicators also involves new information discourse. Work towards sustainability indicators seeks to use information and data drawn from a wide range of other topics (e.g. pollution,
population, resource use) to provide a statistical picture that can
guide policy in a new direction, as in the World Resources Institute
publications. It is argued that the broader quantitative information
results in a broader set of norms for policy. Information and policy
practice are "stretched" to take on board the implications of the new
concept discourse.

The effect is to extend the new concept discourse into the realm of
new practices and specified proposals. Generally, the policy practice
implications of the sustainability concepts and the associated new
information demands and resources are worked out in related discussions on global warming, and so on. And this involves linking concepts of sustainability to specific applications. As Egunjobi (1993:
34) says, the relationship of environment and development is so
complex that "it has always been difficult to give an operational content to the concept of sustainable development or to evolve practical
policy guidelines". Box 3.5 indicates how arguments are employed in
linking sustainability and sustainable development to specific decisions. Sustainability or sustainable development are not the *subject*;
but rather, the elements of the situation are being weighed up using
sustainability concepts.

The first two quotes are concerned with physical measurement,
and the appropriate units to use. The arguments concern quantities
of resource extracted, and quantities of soil exhausted. The second
two quotes are about economic measurement, and again the issue is

---

BOX 3.5   New practice discourse

Theme: measured progress
Topic: sustainability and sustainable development

Limits to exploitation   "The question of the ratio$_{NID}$ of mined sand to remaining resource is of the utmost importance with respect to sustainability . . ." (Hilton 1994: 823)

Valid practice   "Agenda 21 placed a lot of emphasis on re-establishing sustainable livelihoods for the 1–2 billion who are impoverished$_{NID}$ by the lack of water, land cover, soil fertility and fuelwood . . . Sustainability quotient (SQ): the fraction of net farm income per hectare of all land including fallow that is *not* obtained at the expense of extracting soil nutrients." (Stocking 1995: 223, 238)

Accountable value   ". . . Value may be determined in terms of sustainable development: . . . An accounting framework would suggest taking account of changes in the stock of an environmental asset, just as conventional accounting takes into account changes in stocks held by a firm . . . There has been a large body of work on the evaluation of landscape quality . . . Part of the problem lies in the difficulty of defining a unit for consumption or measurement . . . Defining the quantum of landscape . . ." (Willis & Benson 1993: 270, 284–5)

Asset reduction   "The concepts which underpin the economic definitions of sustainable development are those of *natural capital* . . . and of *intergenerational equity* . . . This reduction in the stock of natural capital would not appear in the net product of Zimbabwe as traditionally measured$_{NPD}$ . . ." (Adger 1993: 338, 349)

---

how to determine the units of valuation. Each quote is a challenge, a challenge to existing criteria: new units are required, new scales of value. In this way the link is made between new concept and new information: the concepts demand a new set of measurements, and the effect will be to generate a whole new array of facts, facts that could not come into existence without the new scale adopted. So, here we see how new concept discourse actively promotes new information, by redefining the very language applied to acts of measurement. And the outcome is a new practice: of mining, of soil husbandry, of landscape management, of national economic accounting.

Sustainable development, sustainability

The attempt to create indicators, set associated policy targets and revise current policy decisions is therefore significant because it emphasizes the practicality of these new concepts of sustainability and sustainable development. These are new concepts that are useful in the policy realm, not "just" the product of philosophizing and conjecture. And, in making this point, sustainability is seeking to connect the new environmental discourses of concept, information and practice.

Differences and realignments

Mention has already been made of the ambiguities inherent in the use of the terms "sustainable" and, particularly, "sustainable development". These ambiguities are to be found in the common root "sustain". To "sustain" has many uses: to keep alive, to maintain, to receive an injury. An early usage was legal as in "the appeal was sustained". "Sustainable" is therefore a widely used term, and not necessarily technical; in the OED, its senses are:

1. Capable of being borne or endured; supportable.
2. Capable of being upheld or defended; maintainable.

But both "sustainable development" and "sustainability" are more varied than is suggested by the usual definitional reference-points; they are culturally creative, and also ambiguous. "Sustainability" and "sustainable" now apply to many situations: economically sustainable, culturally sustainable, environmentally sustainable, and each phrase signals a claim to new thinking.

This variety suggests a widespread and diverse influence; but also underlines the divergence of meaning. The differences on sustainability and, particularly, sustainable development can be reorganized in many ways, allowing new alignments to develop. One important structuring of these differences is the use of the weak–strong spectrum of sustainability, which is related to different analytic positions on sustainable development.

The weak–strong sustainability spectrum can be found in many analyses, although mainly from within the discipline of economics (Pearce 1993, Jacobs 1991, Turner 1993a). Strong sustainability "places a premium on the peace of mind in knowing that one is living in a truly sustainable world" (O'Riordan 1995: 22) and involves full, precautionary protection of ecosystems and maintenance of all "natural capital". Weak sustainability seeks rather to preserve "critical natural capital", which maintains life systems, and very weak sustainability looks only to the overall stock of capital, natural and man-made, and allows a degree of substitution between categories to maintain the overall stock level.

These different definitions imply different attitudes towards future generations in terms of the type of world they should inherit, the particular combination of natural and man-made assets that will be available, and also different preferences for taking risks with Earth's life-support systems. These are differences at quite fundamental levels, which require choices to be made before the policy details of achieving sustainability can be tackled.

But these choices are generally mediated through the disciplinary structures of economics, so that the key issue becomes one of relating economic development to environmental protection. In this way, different perspectives on sustainability become linked to different perspectives on sustainable development. Table 3.2 sets out four key positions on sustainable development, specifying them in terms of views on economic development, environmental protection and social or equity considerations. Broadly, the different positions are transformations and elaborations of the weak–strong sustainability spectrum: the quasi-cornucopian position is perhaps off the end of the spectrum, in that it claims we are already in a weak sustainability position; the social choice (as represented by the London Environmental Economics school of David Pearce and his colleagues) explicitly adopts a weakish sustainability criterion, whereas the new economics and limits to growth positions tend towards stronger sustainability criteria. "Strength" is not always perceived or presented as a virtue in these discussions; rather a "weak" position may be held as a mark of reasonableness, practicality and serious policy intent.

Adopting different strengths of sustainability also implies different connections being made between development, the environment and equity. At issue are alternative analyses of the way in which economic processes interact with environmental processes, the extent to which conflicts between these processes exist, and the scope for resolving these conflicts.

For example, the social choice perspective sees a degree of conflict between environmental protection and economic goals as inevitable, but optimistically argues that trade-offs between these two goals can be managed so as to achieve an optimal balance and an overall

Table 3.2 Perspectives on sustainable development.

Approach	Quasi-cornucopian	Social choice	New economics	Limits to growth
Development	Current growth pattern	Marginal change	Substantial change	No growth
Environmental protection	Positive feedback possible	Trade-offs necessary	Negative feedback currently	Absolute limits to growth
Equity	Redistribution possible	Redistribution depends on growth	Redistribution a prerequisite for sustainability	Redistribution possible now
Key figures	Schmitt-Heiney	Pearce	Ekins	Meadows Daly
Sustainability spectrum	-------Weak---Strong-------------------			

maximum level of welfare. Once this has been achieved, then equity concerns can be met by redistribution from within the resulting maximized pool of welfare. By contrast, in the new economics approach, here the distributional pattern of resources is seen as fundamental to the operating of economic and environmental systems, so that redistribution is a necessary prerequisite for achieving sustainable development. The implication is that this redistribution would involve a substantial redefinition of what is termed economic development, allowing a more community-orientated, bottom-up approach, and resulting in new patterns of production and consumption that would reduce the current conflict between environmental production and economic activity (Apichatvullop & Compton 1993, Tisdell 1993).

The discussions on sustainable development have a degree of rhetorical stability. They comprise compatible modes of discourse from different viewpoints. For instance, although the UK *strategy* is more pro-development, the Pearce *Blueprints* are more emphatic about sustainability, and the Brundtland Report is more radical about global equity; nevertheless, the promotion of the new concept is analogous. Sustainability is the somewhat more unstable spin-off. As an arguable world, it is less controlled and it is harder to decide what is at stake. There is less of a tendency to use "sustainability" to finish off a point and more of a tendency to gesture into a different perspective, another discussion. Nevertheless, again the discourses of cultural innovation are confident, especially in the area of new practices and indicators.

It becomes difficult to adopt the position of the judicious critic in sustainability discussions, to argue against the claims of cultural renewal. Rather, critics of the concepts become placed as outsiders and as extremists, beyond the spectrum or range of accepted viewpoints, and this is reflected in the language that is also often extreme. Thus, Beckerman describes strong sustainability as "morally repugnant" (1994: 203). Or Porritt (1993: 38) points up the potential for "political fudge" offered by the sustainable development concept and asks "sustainable development: panacea, platitude or downright deception?":

> The dishonest and platitudinous definitions from political leaders are dangerous, for they draw a veil over the real state of the Earth today, and postpone the time when we stop playing at being a little bit "cleaner and greener", and undertake with resolution and imagination the profound transformation of our industrial society that is now so urgently required.

Porritt's arguments are angry, and the anger is directed at a presumed consensus, almost a global consensus. Although in fact viewpoints diverge, sustainable development and sustainability act centripetally – they draw voices together – to the dismay of some and the hope of many others.

Reviewing the impact of the concepts, O'Riordan (1993: 37), one of the champions of environmental management based around sustainability and sustainable development goals, acknowledges:

> It is tempting to dismiss the term "sustainable development" as an impossible ideal that serves to mask the continuation of the exploitation and brutality that have characterised much of human endeavour over millennia. In the first edition of this book I argued (O'Riordan 1988) that sustainability was being used as a mediating term to bridge the widening gulf between "developers" and "environmentalists". I also contended that the concept was deliberately vague and inherently self-contradictory so that endless streams of academics and diplomats could spend many comfortable hours trying to define it without success.

In this quote we can see many of the key themes of our analysis coming together: the emphasis on consensus and the argumentation constraints associated with rhetorical strategies to achieve that consensus. The professional self-interest that O'Riordan alludes to is well served by concentrating on the apparent reconciliation of economic development and environment protection in the sustainable development agenda; but, as O'Riordan is also contending, there remains the challenge – to deeper reconciliation, reconciliation achieved between a reasonable practical approach and the broader overtones of the concepts. By comparison with most major environment words, there are secure boundaries around this particular part of the environmental agenda, even though the story is as yet unfinished.

Earth, planet, globe, Gaia

Cultural innovation

Suddenly from behind the rim of the moon, in long, slow-motion moments of immense majesty, there emerges a sparkling blue and white jewel, a light, delicate, sky-blue sphere laced with slowly swirling veils of white, rising gradually like a small pearl in a thick sea of black mystery. It takes more than a moment to fully realize this is the Earth . . . Home. (A much used quote from US astronaut Edgar Mitchell, for example in Girardet 1992: 19)

This image of the Earth, conveyed both in words and in the visual counterpart, must be one of the most widespread and influential of environmental images. And the word "Earth" and its variants of "planet", "world", "globe", and more recently "Gaia", clearly have a central place within environmental arguments.

Gaia is the exemplar of the environmental use of a new concept discourse. It refers to a theoretical position in which the Earth is seen as a living entity, a totality. Overcoming the traditional division between the physical and biological sciences, between geology and botany or zoology, the system is analyzed in terms of feedback loops from one element of the system to another, so that the overall entity constantly regulates itself. In the parable of Daisyworld, which Lovelock tells, the ratio of dark to light daisies adjusts in response to temperature and, in doing so, generates changes in temperature.

Gaia is also a concept that is explicitly holistic and ecocentric; as such, it contrasts interestingly with the essentially anthropocentric discussions of population; non- or anti-species-ist positions on population are very rare. Examples of the new concept discourses on Earth and Gaia are provided by Lovelock.

> I wish only to speak out for Gaia because there are so few who do, compared with the multitudes who speak for the people. (Lovelock 1988: xvii)

> . . . Wherever we find a highly improbable molecular assembly it is probably life or one of its products, and if we find such a distribution to be global in extent then perhaps we are seeing something of Gaia, the largest living creature on Earth. (Lovelock 1979: 34)

> It used to be assumed that the escape of hydrogen atoms from the exosphere gave Earth its oxygen atmosphere. Not only do we now doubt that this process is on a sufficient scale to account for oxygen but we rather suspect that the loss of hydrogen atoms is offset or even counterbalanced by the flux of hydrogen from the Sun. (ibid.: 67)

Gaian argumentation is visionary and even poetic, but it can also be informative, as the last extract suggests. Indeed, the status of Gaia as a new concept is bound up with the scientific basis for its hypotheses and predictions, and considerable effort is going into testing the scientific validity of Gaian propositions (Schneider & Boston 1991, van Breemen 1993). The arguments are, therefore, also using new information discourses to extend scientific knowledge, but, because of the connections with the new concept discourse, this also involves a questioning of the current nature of scientific inquiry itself.

The combination of new concept and new information discourses has long been a feature of environmental texts on "Earth", as the following quote from the seminal work *Silent spring* illustrates:

The "control of nature" is a phrase conceived in arrogance, born of the Neanderthal age of biology and philosophy, when it was supposed that nature exists for the convenience of man. The concepts and practices of applied entomology for the most part date from that Stone Age of science. It is our alarming misfortune that so primitive a science has armed itself with the most modern and terrible weapons, and that in turning them against the insects it has also turned them against the Earth. (Carson 1962: 257)

Linkage also arises from the way in which new information and new concept discourses used to discuss Earth and its variants span across the disciplinary boundaries: from space pictures of the Earth to the sociology of modernity, from international politics to climate change, from spirituality to demography. "Earth" is where sciences overlap, and where visions converge. So, we find discussions and new conceptualizations of the Earth and Earth-issues in physics (Capra 1988), economics (Daly 1992), philosophy (Johnson 1991), political theory (Eckersley 1992), gender studies (Shiva 1988) and social theory (Giddens 1994). This has implications for the overall form of the discussion on Earth.

One particular aspect of the argumentative interaction concerns the role of the different disciplines. Disciplinary rivalry in environmental discussions can be intense, with each discipline claiming special authority to speak on the new concepts. The major initiative is undoubtedly the Gaian proposition and how Gaia has interacted with prevailing conceptions of scientific argument and operated more broadly in the argumentative culture of competitive theories.

Lovelock records how scientists reacted to his Gaian proposition:

The normally placid pages of those cool scientific journals, *Nature* and *Science*, have burnt like an inner city, the conservative defenders of ordered gradual change reacting against a revolution for the right to interpret Darwin's great insight. (Lovelock 1988: 12–13)

And Richard Dawkins, from within Darwinian biology, argues that Gaia has failed to satisfy the entry requirements of biology, since it is contradicted by the defining framework of all biological theories – evolution by natural selection through competition and variation:

... There would have to have been a set of rival Gaias, presumably on different planets. Biospheres that did not develop efficient homeostatic regulation of their planetary atmospheres tended to go extinct. The Universe would have to be full of dead planets whose homeostatic regulation systems had failed, with, dotted around, a handful of successful, well-regulated planets of which Earth is one. (Dawkins 1982: 234–6)

Box 3.6 provides a collection of Gaian and anti-Gaian views to illustrate this controversy within science.

The new concept controversy comes into focus as the Gaians propose, and the critics deny. Gaia is integrated science, beyond physics and chemistry and biology; Gaia is atmospheric chemistry and geological time. The contrast is with evolutionary Darwinism, which the Gaians see as archaic: a static context, within which a competition for supremacy occurs. We should, they argue, re-picture the world as itself evolving, alive and responsive. But the Darwinian reply is that evolution means competition, not synthesis, that natural selection means conflict. The Gaians proclaim integration; the Darwinians see it as confusion, a confusion between popular assumptions and scientific theories.

So, is the Gaian view legitimate, a scientific proposal? The conflict is about how far Gaia should be debated seriously, or where it should be discussed. There is the proposal and the critique; but there is also a conflict about the whole exchange itself. How far does science stretch? Should biology be reconceptualized? Or are the boundaries established by previously successful work? The issue is disciplinarity itself, the social arrangement of ideas and knowledge. As Gaia advances, disciplines shift:

Lovelock (1979, 1988) describes the Earth as a superorganism, Gaia, evolved over geologic time by feedback processes that kept atmosphere, oceans, climate and crust in a state comfortable for life by the action of living organisms. Geochemists now generally agree that the peculiar non-equilibrium chemistry of the Earth's atmosphere is dominated by biotic processes... The approaches of geochemistry, biology, and systems theory are converging. (Van Breemen 1993: 1983)

BOX 3.6   New concept discourse

Theme: Gaia and science   Topic: Earth variations

Gaia is science beyond disciplines   "Scientists are also constrained by the tribal rules of the discipline to which they belong. A physicist would find it hard to do chemistry and a biologist would find physics well-nigh impossible to do . . . As a university scientist I would have found it nearly impossible to do full-time research on the Earth as a living planet." (Lovelock 1988: xiv)

Gaia is new science (integrated) vs old science (Darwinism)   "As a hypothesis, Gaia integrates the evolution of the biota with the well-documented transformations in the surface and atmospheric chemistry of the planet through geologic time. That the Gaia concept cannot be framed by the stilted terminology of neo-Darwinistic population biology is not surprising since Gaia is a hypothesis based in sciences that neo-Darwininism proudly ignores." (Margulis & Hinkle 1991: 15)

Gaia is not science   "The Gaia hypothesis is an extreme form of what . . . I shall continue to call the 'BBC theorem'. The British Broadcasting Corporation is rightly praised for the excellence of its nature photography, and it usually strings these admirable images together with a serious commentary . . . For years the dominant message of these commentaries was one that had been elevated almost to the status of a religion by pop 'ecology'. There was something called the 'balance of nature' . . ." (Dawkins 1982: 236)

Gaia is pseudo-science   "On the Fringe of Rio: Vikings, Vegetarians and Taoists at the Alternative Summit
    They are beating the drums . . . No question, this is the greatest place
    to be on the planet right now if you can't get a ticket to the NBA play-
    offs. It's hot, it's weird, it's where it's at.
The best idea floated so far, and it came right from the stage during the opening ceremony, is to change 'Tuesday' to 'Gaiaday', in reference to the Greek word for the Earth Goddess . . . Let Jay Smith, a 41-year-old schoolteacher from Berkeley, Calif., who made the pilgrimage here, explain:
    Gaia is our planet, it's one big organism, a living thing. It's really important.
    It's one super being that's been living for about 3.2 billion years. We, the
    humans, are the brains of Gaia. We're the first ones that have made Gaia
    self-aware.
So then, does that mean each person is sort of like an individual neuron?
    Yeah, yeah, exactly. Good thought." (*Washington Post*, 2 June 1992)

Here is the Gaian argument with science: it seeks to advance science, by integrating sciences; it also leaps above science, into overt metaphor and world-view.

> Who can argue against the thesis that we happen to live on that exceptionally rare planet where, by sheer luck, life has so far escaped global destruction?" (ibid.: 207)

Lovelock himself makes the double proposition of integrated science and new world vision: "I wish only to speak out for Gaia because there are so few who do, compared with the multitudes who speak for the people." (Lovelock 1988: vii)

Differences and realignments

Broadly, two different viewpoints on Earth may be distinguished: thin Earth and thick Earth, although one can also posit a spectrum based on these viewpoints or a dialectical interplay. These two viewpoints are encapsulated in Ted Hughes's poem "Creation/Four Ages/Flood". First comes thick Earth:

> And the Earth, unbroken by plough or by hoe,
> Piled the table high. Mankind
> Was content to gather the abundance
> Of whatever ripened.
> Blackberry or strawberry, mushroom or truffle,
> Every kind of nut, figs, apples, cherries,
> Apricots and pears, and, ankle deep,
> Acorns under the tree of the Thunderer.
> Spring weather, the airs of spring,
> All year long brought blossom.
> The unworked earth
> Whitened beneath the bowed wealth of the corn.

and then a dismal view of the thin earth:

> Earth's natural plenty no longer sufficed.
> Man tore open the Earth, and rummaged in her bowels.
> Precious ores the Creator had concealed
> As close to hell as possible
> Were dug up . . . (Hughes 1994: 8–9, 10–11)

The thin Earth viewpoint stresses the concrete and practical issues of resource depletion, scarcity and degradation, as well as the practicalities of managing these problems. Here the world is both a static entity and an object, a static object that we have come to realize must be saved. Or rather, as Ingold (1993) points out, it is life on the surface of the world that must be saved. For this view of the planet, the Earth "consists of pure substance, physical matter, presenting an opaque and impenetrable surface of literal reality *upon which* form and meaning is overlaid by the human mind" (ibid.: 37). To know this world is, as Ingold puts it, "a matter not of sensory attunement but of cognitive reconstruction", a task that involves mapping rather than direct engagement.

The thick Earth view is based on a more dynamic concept of Earth as an organism in which ecosystems rather than resources are the focal point. Such an organism is not "saved", it is respected and engaged with. Using a related conceptualization of the Earth as a sphere rather than a globe, Ingold bases this view in a variety of historical and cultural contexts (and see also Albert 1993) and, of course, we have the current development of Gaia as a contemporary example. In this view the Earth is a lifeworld, experienced from the centre, which cannot be separated into "life" plus a physical entity. It cannot be viewed from outside either.

The thin Earth view is prominent in the media when environmental issues are discussed. But the alternative is never far away, the thick Earth, the rich Earth, fathomless as the thin Earth is brittle. Box 3.7 shows how the views are counter-posed, with numbers favouring thin Earth but vivid glimpses of the other world.

The quotes incorporate several of our chosen environment words: population, global warming, pollution, biodiversity/species. The negotiations on the international Biodiversity Convention frame much of the coverage cited here. The environment words cross each other, as the image of the Earth is presented and re-presented. Although the issues are practical, the contrasts are metaphysical. What is "Earth", what does "world" mean, where do "we" belong?

But there are implications in terms of whose viewpoint is being adopted. From his anthropological stance, Ingold argues that "once the world is conceived as a globe [the thin Earth view], it can become an object of appropriation" (Ingold 1993: 39) and that the image of the world as a globe is a colonial one – "it presents us with the idea

---

BOX 3.7   Controversies

Theme: thick Earth and thin Earth
Topic: media bite on Earth variations

Thin Earths

THE POISONED EARTH SURFACE   "Mayor Francisco Villarreal, of the National Action Party, said he was powerless to take action against Presto Lock because only the PRI-controlled federal government [of Mexico] has jurisdiction to sanction companies$_{NPD}$ that emit pollutants into the air or Earth." (*Washington Post*, 24 December 1993)

THE BEREAVED PLANET   "As the human population continues its rapid expansion, up to 100 different plants and animals$_{MID}$ a day may be becoming extinct – the greatest planetary loss in tens of millions of years." (*Independent*, 30 December 1993)

THE VULNERABLE WORLD   "For example, more than 90 per cent$_{MID}$ of people questioned were concerned about the destruction of the world's forests, and more than 80 per cent$_{MID}$ worried by threats of acid rain, global warming, and the destruction of the ozone layer." (*Guardian*, 30 December 1993)

THE DENUDED EARTH   "The majority of the Earth's species$_{MID}$ are found in developing countries – and it is in these countries that they are dying out fastest." (*The European*, 31 December 1993)

Thick Earth

THE SECRET PLANET   "A growing body of scientists now believes$_{MID}$ that a giant, untapped abundant world of life is buried deep beneath the planet surface, extending down for miles with a total mass that could exceed that of all previously known life on Earth." (*Daily Mail*, 29 December 1993)

---

of a preformed surface *waiting to be occupied*, to be colonized first by living things and later by human (usually meaning Western) civilization" (ibid.: 38). But further, and picking up on a repeated thread in environmental arguments, there is the potential for the use of the globe as metaphor to present an apparently "common" viewpoint. Ingold specifically talks of "appropriation for a collective humanity" and points to how this can privilege "globally conscious Westerners" with their ontology of detachment over local views based on an ontology of engagement – the thick view, the sphere. Here, different visions of the Earth are in engagement with direct implications.

Ingold is presenting a critique of the use of a globalizing viewpoint, one that is premised on the existence of conflicts at a global scale, although experienced more locally. It is a critique that mourns the loss of local identity on the basis of reactions to globalized problems. This is a critique echoed by other analysts (Shiva et al. 1991). But a more positive characterization of the globalizing viewpoint can be found, particularly in documents addressed to policy-makers, such as the Brundtland Report and Al Gore (1992). Here the emphasis is on common interests as expressed through a co-operative international community, which, using communication rather than more coercive methods (see also Gordon 1993), can build on diversity and not suppress it.

So the two viewpoints of thick and thin Earth are crossed by that prevailing cleavage in environmental debates between the perception of community or of inherent conflict. The structure of viewpoints means that there are cacophonies of different voices about the Earth, the planet, the world. There is the central dispute over whether global problems imply global common communities of interest and response; the dispute between, for example, Gore or Brundtland and de la Court, between de la Court and Lewis. And there is the profound philosophical difference over whether the Earth is a biological system, an organism, or merely a crust on which people, a uniquely privileged species, have their special existence.

The environet as an open system

The environet *is* a system. Each topic is tied to the other topics; each argument interacts with the others; all environment words are interrelated. When the system changes in one place, the change impacts elsewhere. So, in this chapter, arguments about technology or about common interest overlap topics, they connect pollution, global warming, sustainable development and Gaia. But the environet is an *open* system: the connections are fluid, multiple and ambiguous. The topics are mutable as well as related: they merge, overlap and separate. The edges are ambiguous. Global warming arguments both are and are not extensions of pollution arguments; the Gaian Earth is both welcoming and indifferent to sustainable development.

The environet is open, and, above all, it is *cultural*, a system of argued meanings. To analyze this system, we must take up the position of an interpreter, not a detached outsider. Environment's meanings are matters of interpretation, not "observation". But interpretation is always arguable, there are always other possible meanings, even to single quotation, text or argument, let alone to a whole cultural system. Rhetoric is about plurality, the many meanings we experience from the interpreter's position. For instance, the environet is *both* about a new sense of limitation, and about new horizons: this ambiguity radiates out from sustainable development to other topics, such as resources, population and Earth.

The environet carries meanings, and these meanings are often profound. For instance, modernity is read and re-read by the varying arguments. Pollution casts modernity as villain and also as rescuer. Global warming is a modern problem; it also invites further modernization. The arguments sway to and fro, modernity is both dismaying and hopeful. Sustainable development may modernize management and measurement; it can also argue with modern society.

The environet leaves no fixed points of reference. Nature is also open to different interpretations. Global warming initiates arguments about nature, nature as a physical system but complex and responsive, and a potential object for new management. Gaia also starts arguments about nature's sensitivity, but the physical system is now also almost a conscious system. This provokes new arguments, on the redefinition of science itself.

We are becoming used to uncertainty and probability in scientific knowledge; indeed they are exemplified in global warming or resource arguments. Rhetoric extends uncertainty and probability to cultural knowledge. Environmental topics are unpredictable. We watch the viewpoints competing, the discourses evolving. Certain developments are evident: the enriching of sustainable development by new practice arguments, the challenge between Gaian and Darwinian paradigms. However, the outcome is uncertain. Rhetoric reveals possibilities. For the patterns of cultural argumentation are spontaneous. We create them. As we analyze further the *processes* of cultural argumentation, we gain a deeper understanding of how the arguments can be both patterned and unpredictable.

CHAPTER FOUR

Environmental ethos

Environmental arguments are cultural arguments: culture is the subject and it is also the context. No-one wins a cultural argument by pure logic: to persuade one must appeal, be an appealing kind of person, the right kind of character for the argument. The texts convey ideas, and they also create personae, rhetorical characters. So, having mapped the topics of environmental argument, we begin the exploration of forms of argumentation with an ethos analysis. This kind of analysis represents the drama of voices as these environment words interact with the culture. In the drama, different personalities emerge, different textual personae associated with the key words. The words encourage typical attitudes, they bring their own colouring to an argument. The ethos analysis illustrates a central theme of environmental rhetoric, the human dimension, the ways in which the concepts *characterize* arguments. There are many environmental personalities and voices, with different impacts in argument. The analysis examines how personal quality is invoked in order to create a certain motivation in the audience and thereby to persuade (Burke 1950, Nussbaum 1990, Billig 1992). As such, the ethos of environmental arguments plays a central role, for it links strongly to notions of authority, legitimation and the criteria by which the argument as a whole is to be judged (Putnam 1981, Nussbaum 1990).

There is also an ethical drive in invoking ethos, to redefine culture itself:

Conservation is getting nowhere because it is incompatible with our Abrahamic conception of land. We abuse land because we regard it as a commodity belonging to us. When we see land as

119

a community to which we belong, we may begin to use it with love and respect. (Leopold 1949: 223–4, in Wall 1994: 76).

Environmental argument continues as a process of reviving such ethos, a revival that at the same time generates new ideas, new facts, and new values.

We explore the role of ethos in environmental argument by looking at the variety of appeals through self-representation: vision, farsightedness, realism, reverence, and objectivity. We specify variants and consider the development of counter-ethos. Picking up on the topic analysis in preceding chapters, the quotes in this and succeeding chapters are annotated to indicate the use of the various discourses we have identified alongside other indicators of the developing rhetorical analysis. This provides an increasingly rich visual account of our own arguments.

Essential vision: sustainable development, sustain, sustainability

There is a common ethos of sustainable development texts, an ethos that we have termed essential vision. The most direct expressions are in the Mugabe and Gayoom texts in Box 4.1: "vision of a common future", "hope must be sustained". This ethos seeks to combine two takes on the temporal dimension of environmental concerns. On the one hand, the pursuit of sustainability or sustainable development is essential; there is an urgency implied and a desire to convey a motivation for acting *now*.

We need a sense of urgency as well as a realization of the need for global involvement. Urgency because time is running out. And global involvement because without it the efforts of individual nations will be undermined by those who refuse to shoulder their proper burdens. (John Gummer, Secretary of State for the Environment in the introduction in UK Government 1994)

This aspect is particularly well illustrated in the extracts from the central text, the Brundtland Report, in Box 4.1. The other aspect of

BOX 4.1   Ethos

Theme: essential vision   Topic: sustainable development

Death and birth   "If we do not stem the proliferation of the world's deadliest weapons, no democracy can feel secure. If we do not strengthen the capacity$_{NPD}$ to resolve conflicts among and in nations, those conflicts will smother the birth of free institutions, threaten the development of entire regions and continue to take innocent lives. If we do not nurture our people and our planet$_{NCD}$ through sustainable development, we will deepen conflict and waste the very wonders that make our efforts worth doing." (President Clinton, reported by Associated Press, 27 September 1993)

The world without foundations   "In many parts of the world, the population is growing$_{MID}$ at rates that cannot be sustained by available environmental resources . . ." (WCED 1987: 11)

The vista of progress   "A safe, environmentally sound and economically viable energy pathway$_{NPD}$ that will sustain human progress into the distant future is clearly imperative. It is also possible." (ibid.: 15)

The human future   "Humanity has the ability to make development sustainable$_{NCD}$ – to ensure that it meets the needs of the present without compromising the ability of future generations to meet their own needs." (ibid.: 8)

A principled order   "In its report *Our common future* (WCED 1987), the Brundtland Commission drew attention to the monumental political and institutional changes$_{NPD}$ that would need to be made if the principle of sustainable development was to be implemented at all scales, from the local to the international." (ibid.: 183)

Total culture   "Full sustainable development involves a cultural shift,$_{NCD}$ not just economic and political tinkering." (Pearce 1993: 185)

New unification   "Preparing strategies$_{NPD}$ for sustainable global development and its impact on the environment is a daunting task. It calls for imagination, determination and above all a vision of a common future." (Mugabe 1991: 10)

Reason and absurdity   "The hope must be sustained and realized. In the face of a global threat, anything less than an all-encompassing international commitment and effort can become futile in this colossal struggle." (Gayoom 1991: 17) the ethos of essential vision relates to futurity, to the underlying nature of the concepts as concerned with future as well as current generations. Here the ethos encourages the audience to empathize with future generations. This is a projection, and it contrasts, as we shall see, with population arguments, for instance, where the ethos conveys the feeling that the problem, although expressed in future trends, is with us *now* and with global warming, where the ethos implies seeing the problem on the horizon although implying some connection with current actions.

The futurity of sustainable development arguments also links into a spiritual dimension. This spirituality is exemplified in Arne Naess's statements on deep ecology, the attempt to fashion a new philosophy and political movement around the environmental agenda. Naess (1994: 1) argues for "an ecosophically sustainable development":

> From the point of view of the history of ideas one may trace a line of thinking roughly suggested by the formula "from the idea of "progress" to that of "development" and "economic growth", and from these ideas to that of "ecologically sustainable development"." What some of us hope for is a further step along this line from ecological to long range "ecosophical" development with emphasis on the need for wisdom (*sophia*) as much as need for science and technology. If this line is followed we shall have to study the loss of beliefs and cultural identity now happening under the tremendous impact of the economy and technology of the large, powerful and rich industrial societies. As members of these societies we are largely responsible for this impact and the resulting cultural shock. Any model of ecologically sustainable development must contain answers, however tentative, as to how to avoid contributing to thoughtless destruction of cultures, and to the dissemination of the belief in a glorious, meaningless life.

Naess connects the spiritual dimension of sustainable development with the Gaia hypothesis (at least in Northern contexts):

> People try to protect a vital need for meaning and what is necessary to maintain that meaning. A development that is expected to reduce that meaning is not sustainable. We are led to a concept

of sustainable development for this satisfaction of human needs which protects the planet also for its own sake. The Gaia hypothesis has shown its value not only as a working hypothesis, but also as a way of experiencing the Earth as alive in a broad sense adapted to people in cultures with Western science. (ibid.: 5)

The spiritual dimension of essential vision is also given expression through the concept of stewardship. The precise way that stewardship is expressed varies with the cultural context. For example, "stewardship" may be overtly theological (Gore), or more moral (Gummer) or a combination (Mugabe):

In the Judeo-Christian tradition, the biblical concept of dominion is quite different from the concept of domination, and the difference is crucial. Specifically, followers of this tradition are charged with the duty of stewardship, because the same biblical passage that grants them "dominion" also requires them to "care for" the earth even as they "work" it. The requirements of stewardship and its grant of dominion are not in conflict; in recognizing the sacredness of creation, believers are called upon to remember that even as they "till" the Earth they must also "keep" it. (Gore 1993: 243)

This document is only the beginning. Year by year we shall need to revise and refine our policies so that our economy can grow in a way which does not cheat on our children. (Gummer in the introduction in UK Government 1994)

We inherited the Earth from our forefathers and hold it in trust for our children. It is a debt of honour that we should pass it on in a livable state – at the very least the state in which it was passed on to us. (Mugabe 1991: 10)

The ethos of sustainability discussions and the nature of the new concept discourse being used, therefore, emphasize the spiritual dimensions. But there is another dimension to the discussions, a dimension that stresses the practicality, the usefulness of these concepts – as in the discussion of indicators and targets we identified when considering the use of new practice discourse in relation to this topic. For some, the spiritual dimension of sustainable development, making rich use of new concept discourses, is clearly threatened by the attempt to shift into new practice discourses through the use of new indicators, unless the indicators themselves are radically transformed:

> We, the rich, are poor in deep satisfactions requiring simple means, the means being material, spiritual, or beyond that somewhat arbitrary distinction. The term "life quality" is used as in that part of life quality research where the life quality indicators clearly are distinguished from the material or rather, means-orientated indicators of established standard of life statistics. (Naess 1994: 9)

> If an assault on growth is one half of the Green economic agenda, sustainability, providing an alternative to constant expansion, completes the whole. Economic systems should be infinitely sustainable, cyclical in nature and able to recycle energy and resource inputs. Rather than being based on quantitative measures of gross national product, their goals should be ecologically centred and qualitative. (Wall 1994: 126)

The arguments invoke a heightened vision, a transcendent view. But for others it is an extension of the argument to move from specifying the concepts to turning them into policy tools. Thus, although the Brundtland Report is most concerned to make the basic switch in thinking involved with new concept discourses, the Pearce *Blueprints* (1989, 1993, 1995) seek to go beyond concepts and argue for the way in which these goals can be realized. Hence, the *Blueprints* particularly invoke the envisioning aspect of the ethos, not to transcendal effect but to demonstrate that such a vision can be created and made real, that it is *reasonable* to make such a vision an "essential" element of political discussion.

> Translated into realisable political action, sustainable development is more about changes of emphasis than a wholesale restructuring of decision-making. (Pearce 1993: 10)

Technical vision: energy

A different form of vision is provided by the ethos of energy arguments (Box 4.2). This is a technical vision, with an emphasis on scenarios that can be chosen between and made real. This is a vision that is promised soon. There are benefits to the power industry, the pipeline technology is available, "new tricks" are here.

These invocations of a technical vision promise us a low energy scenario (Alcamo & De Vries 1992) in a new era. From within

---

BOX 4.2   Ethos

Theme: technical vision   Topic: energy

Technological spread   "Clearly, the greater the compression that can be achieved in the compressor stage and the higher the gas temperature that can be tolerated by the turbine stage, the greater the potential efficiency of energy conversion. These have been precisely the goals of the aero-engine manufacturers over the past four decades. The power industry is now able to benefit from them."ₙₚₒ (Chester 1993: 48)

Technological revolution   "This revolution in the [natural gas] industry also required the laying of a greatly expanded length of high-pressure pipeline operating at up to 75 bar.ₘᵢₒ This now provides a national energy transmission systemₙₚₒ which continues to function effectively through hurricanes and blizzards when other forms of energy supply have broken down." (Rooke 1993: 73)

Technological new era/horizon   ". . . The need for a truly global, comprehensive and strategic approachₙᴄₒ to the energy problem . . . Within the context of the SEI [Strategic Environment Initiative], it ought to be possible to establish a coordinated global programₙₚₒ to accomplish the strategic goal of completely eliminating the internal combustion engine over, say, a twenty-five year period." (Gore 1993: 325–6)

Technology in another nature   "New tricksₙₚₒ like these will use the Sun's energy – there is no other – but in ways far more elegant than those we have deployed so far. Oil is only what bugs have made of rotting forests. Coal is even less clever stuff: just anaerobically decomposed and compressed forest. Modern technologiesₙₚₒ promise vastly to improve on them." (North 1995: 76)

---

philosophy we find a critique of this necessity to move urgently into new futures

> Having been stamped by the *Gestalt* of the atomic age, humanity has been compelled to unleash the monstrous energies of the atom; and then compelled to develop ever greater methods for its control: "For the self-establishing of life, however, it must constantly secure itself anew." (Zimmerman 1990: 198–9, paraphrasing and quoting Heidegger)

Other voices stress the need for a policy perspective to make the future scenario real. Flavin & Lenssen (1992) claim that the fossil fuel age may be nearing its close but that policy reform is needed to make the transition to a renewable energy economy. Gore also makes this point (see Box 4.2).

But such a vision needs defending; can the technology really deliver?

> Despite the broad public consensus that renewable energy is to be welcomed and is preferable to other forms of power generation, innate conservatism and a degree of ignorance about the realities of renewable energy combine to generate scepticism and, sometimes, hostility towards renewables technologies and individual projects. (Rae 1993: 101)

This is particularly necessary when the vision does seem to take us to the outer limits of the possible. So, in Johnson-Freese et al. (1992) we have the prospect of mining the Moon for new energy sources, a vision indeed of a technologically new world.

Systematic far-sightedness: resources

Ethos is important in all environmental arguments, but in resource arguments ethos is supreme. On the one hand, there are inherited tones and voices, the voices of the long debate, Malthusian and counter-Malthusian presences in discussion. On the other hand, there are personalities that carry authority now, and in particular personalities

BOX 4.3   Ethos

Theme: systematic far-sightedness
Topic: resources

Malthusian systematic prudence   "Even if the problems of distribution were solved entirely to their satisfaction (for they are absolutely right to point out that a much more equitable distribution of the Earth's resources is just as much of a problem as population itself), we would still face desperate ecological problems as a consequence of there simply being too many people asking too much of a finite resource base." (Porritt 1993: 28)

Far-sighted calculations   "In a didactically superb manner he [Professor Jay Forrester] demonstrated, <u>by means of quickly sketched flow diagrams,</u>$_{NCD}$ how population, industry and resources interacted with one another, and in this way he succeeded in restoring our confidence that a quantitative model depicting the Predicament of Mankind, if not in its entirety, at least in some of its important aspects, could actually be built." (Pestel 1989: 23)

Far-sighted strategic planning   "<u>Deliberately limiting growth</u>$_{NCD}$ would be difficult, but not impossible. The way to proceed is clear, and the necessary steps, although they are new ones for human society, are well within human capabilities. Man possesses, for a small moment in his history, the most powerful combination of knowledge, tools, and resources the world has ever known." (ibid.: 184–5)

Far-sighted strategic planning   "If sustainable development is to be achieved, <u>we will have to devise institutions,</u>$_{NPD}$ at all levels of government, to reallocate the use of stock resources towards the future, . . ." (Norgaard 1992: 84)

Far-sighted science   "<u>The discovery</u>$_{NID}$ of uranium was new knowledge that increased our resource base. <u>The subsequent discovery</u>$_{NID}$ of the dangers of radioactivity did not further expand the resource base, but contracted it." (Daly 1992: 185)

Far-sighted social theory   "*individual* freedom in the market . . . <u>It is hopelessly outdated</u>$_{NCD}$ where *common* resources must be shared, i.e., air, water, or open space, in today's crowded urbanized societies, where win–win rules are needed for equitable access." (Henderson 1993: 98) that are believable when the future is being discussed. The traditional arguments were always rather systematic, from the metaphors of the reservoir onwards. But, as Box 4.3 illustrates, foresight has implied a wide variety of cultural personae over the two modern centuries: wisdom, prudence, caution, audacity. The contemporary discourses are technical, with a high degree of complexity and elaboration, drawing on economics, science and social theory. At the same time, they seek general authority. The result is an ethos of systematic far-sightedness. Even where viewpoints differ, they are rivals for the same kind of authority, and so they tend to imply the same range of ethos.

On one side, we have iron law voices, Porritt and Daly, in this instance; these are Malthusian Utopias coping with the iron laws of decline and diminution. Other voices are Utopian, directly idealistic, notably Henderson. But far-sightedness is about limits – we foresee the dangers with Daly or Porritt – and it is also about new opportunities, new rules, new ideals. Foresight can be used to plan, like Norgaard, or it can be used to prophesy, like Henderson.

The passages project advanced thinking, advanced systems of thought: "a quantitative model", "rules are needed". The voices are combatively up-to-date, difficult, overwhelming even, as in the depiction of Forrester motivating the Club of Rome with "a didactically superb manner". This is top-down argumentation, not empathetic or achieved via experience, but transcendent thinking: the ethos of intellectual innovators. The strategy is a scene-stealing one: the other side is always caught out, too old, not in touch. The argumentation polarizes in the form: our side is ahead of its time versus their side is out of date. The viewpoints conflict, but together they keep alive "modernity's consciousness of time open toward the future" (Habermas 1987: 12). Contemporary knowledge on resources is systematic, but the systems diverge and contrast. There is not one meta-system, no emergent system to synthesize all the rivals.

Inclusive far-sightedness: global warming

In the case of climate change and global warming arguments, the far-sighted ethos is produced by using phrases of the new information

BOX 4.4   Ethos

Theme: inclusive far-sightedness   Topics: global warming

Scientific perspective   "This principle, known as the greenhouse effect, explains why_{NID} gases produced by human activity will probably cause the Earth's average temperature to increase within the lifetimes of most people living today." (NAS 1990: 63)

Social perspective   "The burning of fossil fuels puts into the atmosphere carbon dioxide, which is causing gradual global warming. This 'greenhouse effect' may by early next century have increased average global temperatures enough_{NID} to shift agricultural production areas, raise sea levels to flood coastal cities, and disrupt national economies." (WCED 1987: 2–3)

Techno-social perspective   "The use of energy accounts for a major fraction of all anthropogenic emissions_{NID} of greenhouse gases (Intergovernmental Panel on Climate Change, 1990), and in most industrialised countries the use of transportation fuels and electricity accounts for a major fraction of all energy-related emissions." (DeLuchi 1993: 187)

Techno-policy perspective   ". . . Consider the types of uncertainty that exist. [Box 1.3 summarizes] the types of gases contributing to the "greenhouse effect", and the relative contributions of each type, . . . It is clear that_{NID} the future warming, beyond that to which the Earth is already committed, will arise mainly from CO_2 and CFC emissions. It follows that part of the uncertainty_{NID} about global warming will arise from uncertainty_{NID} about the future fuel mix." (Pearce 1989: 13–14)

Political perspective   "After months of wrangling, the European Union has agreed to ratify a UN treaty on combating global warming . . . In theory this joint action_{NPD} by such a large group of advanced industrial nations should mark an important advance to combat the threat of climate change in the next century. But in practice the deal struck in the small hours has elements of fudge and uncertainty in it." (*Independent*, 17 December 1993) discourse, in particular phrases relating to uncertainty and also phrases of timescale, whereby there is a shading of the present into the future.

The first two quotes in Box 4.4 are from major institutional texts, from the US National Academy of Sciences and the WCED, and they are representative of an approach that feels authoritative and definitive, although speaking the language of uncertainty. By contrast, the Pearce quote is exploratory: what is the nature of the newness of new information discourses, how do we treat uncertainty as information? And the last quote, from the heightened context of the media, is critical: it looks ahead over the heads of the establishment and asks whether their foresight is real or spurious. In doing so it speaks on behalf of common interest, defined over and against the authorities. Varied though they are, the texts each embody a characteristic far-sighted ethos, the searching gaze, the vision over the horizon.

Although this claim of far-sightedness is the prevailing ethos supporting global warming texts, shadows of other forms of ethos can be found, as with any complex text. Another presence is emotive objectivity (the dominant form in population arguments), as in:

> As the poles warm up at a faster rate than the equator, the difference in temperatures between them lessens, as does the amount of heat that must be transferred. As a result, the artificial global warming we are causing threatens far more than a few degrees added to average temperatures: it threatens to destroy the climate equilibrium we have known for the entire history of human civilization. (Gore 1993: 98)

And there is, by contrast, reassuring realism (which we will link to pollution arguments in general):

> Bad news sells newspapers and swells television audiences, good news does not. It was therefore inevitable that the media, over the past decade, should have given greater emphasis to the dramatic aspects of environmental change – global warming, ozone holes, rising oceans, a nuclear winter – than to solid and sometimes difficult scientific evidence that puts these and other environmental problems into a proper perspective . . . Given the rate at which computer science is advancing . . . More sharply

focused predictions of greenhouse warming, and consequential rainfall patterns and ocean levels will then be possible. (Cartledge 1992: 1–3)

But the very phrasing of these arguments makes it clear that the reassurance, the counter to more emotive language, depends on the claims of scientists and policy-makers to achieve a far-sightedness based on both the use of scientific expertise and the ability to negotiate successfully between parties.

The ethos of inclusive far-sightedness arises in the context of global warming because of certain difficult requirements. On the one hand there is the definition of the issue in terms of the knowledge of what is happening, articulated in a discourse of new information in which uncertainty over that information is a key feature associated with the new forms of scientific rationality. On the other hand, there is the concern with distributional aspects of the problem, resulting in a discourse of new practices based around negotiation and linked in to recognition and perception of the distinction between collective benefits and individual costs. The ethos of global warming arguments therefore has to straddle these tensions in order to convey authority and claim legitimacy.

One crucial requirement is to convey the expertise of science; the ethos reinforces the scientific basis of most new information discourses on this topic and gives prominence of place in the arguments to the development of scientific models and the discussion of their outputs. This applies not only to models concerning climate and physical processes but also to the use of economic models and the attempt to integrate climate and economic models (Duraiappah 1993, Nordhaus 1993). This type of argument places an emphasis on the output of the models and "scientific" arguments. Policy arguments are also assessed in terms of their impacts, whether on CO_2 or an economic indicator such as GDP (e.g. Barker et al. 1993, Blitzer et al. 1993, Manne & Richels 1993). Inter-country comparisons, which are common, also use the variable impact of policy in terms of these indicators as their basis of comparison.

Another requirement arises from the distributional aspects of global warming, underpinned by new practice discourses of negotiation. There is an appeal to common interests as a strategy for reinforcing arguments and defusing tensions: "It is in the interest of all the world that climatic changes are understood . . ." (Gayoom 1991: 16). So, there is an emphasis on the ability to seek out common interests, to present the global warming problem as "our" problem, with an associated emphasis on a common solution, as in carbon taxes or permits. The search for a diplomatic solution depends on finding common ground, persuading parties of their common or collective interest, privileging the "our" over the "mine".

There are clear links here with scientific expertise. For example, the development of overarching scientific models applicable to different countries and scenarios suggests the underlying commonality of the problem, its diagnosis and treatment. Invoking science itself is a profound way of seeking such commonality (Taylor & Buttel 1992). Scientific expertise does not completely defuse the tension between the collective and the individual. The way the texts seek to offer a solution to the tensions implicit in their arguments is by combining the authority of the scientific expert and the negotiator of the common good in the ethos of inclusive far-sightedness. This ethos suggests an ability to look beyond the moment to consider the future, to look beyond the partial to consider all parties, to resolve the tension of the individual versus the collective.

Reassuring realism: pollution

A common ethos of official pollution texts involves the presentation of a persona that is both reassuring and realistic; even texts that are critical of current regulatory practice will involve calls for measures that will reassure and emphasize that these calls are realistic (Rose 1990). Some illustrations are provided in Box 4.5, taken from official UK government documents: the UK *strategy for sustainable development* (1994) and reports from the Royal Commission on Environmental Pollution. Indeed, the scientific, or perhaps technical, language of thresholds, levels of toxicity and the statistics of epidemiology provide support for such a reassurance. Scientific reassurance is seen at work in the first extract from the UK *strategy*, where the stress falls on scientific measurement, on the detail, and on research. The argument by-passes the individual perception of experience. Yet the other view, that there are invisible long-term risks that

BOX 4.5   Ethos

Theme: reassuring realism   Topic: official texts on pollution

Low-level problems   "In the UK, acute health problems are now rarely the result of environmental causes, and attention has shifted to the health implications[NID] of long-term, low-level exposure to environmental pollution. Cause and effect are harder to assess here and extensive studies may be needed in such cases to establish what the dangers are." (UK Government 1994: 7)

Rare problems   "Major pollution incidents from sewage and agriculture are reducing rapidly. But a small proportion of rivers continue to be affected by severe industrial pollution."[NID] (ibid.: 9)

Localized problems   "Pollution in the atmosphere, agricultural practices and contamination from industrial plants or from wastes can all affect soil quality.[NID] Fortunately, these problems are not widespread in the UK, although there are particular concerns in some areas." (ibid.: 10)

Minimized problems   "The main objectives of IPC are:
(a) to prevent or minimise[NPD] the release of prescribed substances and to render harmless any such substances which are released; and . . .
(g) to maintain public[NPD] confidence in the regulatory system through a clear and transparent system that is accessible and easy to understand and is clear and simple in operation;" (Department of the Environment/Welsh Office 1993: 3)

Reduction of problems (regulation)   "BPEO as we describe it is a procedure that would lead, if properly implemented,[NPD] to reductions in environmental pollution and to improvements in the quality of the environment as a whole." (RCEP 1988: 2)

Reduction of problems (technology)   "Technology is already helping[NPD] to improve air quality; we advocate further advances here. We have also sought to stimulate the manufacture and sale of more fuel-efficient vehicles, and the development of appropriate technologies for improved traffic management."[NPD] (RCEP 1994: preface)

Note:
IPC – integrated pollution control
BPEO – best practicable environmental option
RCEP – Royal Commission on Environmental Pollution

ordinary care cannot circumvent, is also palpable inside the text. Hence, there is realism tempering or undermining the reassurance.

The second extract from the UK *strategy* illustrates another rhetorical strategy for dealing with the implicit threat of urgency in pollution incidents. Here, several shifts are promoted: from "incidents" to "affected", from event to process, from acute to chronic. The rhetorical strategy is redefinition, it is argumentatively reinterpreting the way "pollution" is understood, away from the spectacular event towards the background, away from disaster towards risk, and away from high risk towards medium or low risk. Cumulatively, the effect is scene shifting, creating new contexts for debating pollution, new sciences, and new expertise.

For the basis of the ethos of realistic reassurance must lie in the demonstration that past pollution has been dealt with and therefore so will current and, by implication, especially future pollution. A problem for this ethos of reassurance is the continued voice of disclosure, of the provocateur or environmental watchdog and there is also the problem of the evidence of past contamination, which suggests that all such problems have not been adequately dealt with in the past (Hird 1993). This ethos of reassuring realism interacts with the new information discourse of disclosure, usually to close it off and build instead on the new practice discourses to suggest that pollution can be dealt with. In its overall effect it can, therefore, tend to deny the ethical position on pollution in favour of a more technocratic position.

Reverent objectivity: Earth

The discussion of "Earth", and its variants such as "globe" and "planet", relates to many different visions, some invoking a thick Earth, some a thin Earth, some common global interests, some local identity. The diversity of visions results in many different bases of claims to authority being used, but there is a common ethos that underpins these different precise claims, and this can be termed reverent objectivity.

As in so many other aspects of environmental argument, there is a need to draw on objective reference points, but this need not mean scientific knowledge as constructed in the North. Objectivity results from being correctly placed to perceive and understand the planet. For those in the North, this has often meant the ability to stand (or orbit) physically outside the Earth and see it as a global whole, or to understand it through a particular set of scientifically accredited relationships. For those adopting a thicker Earth perspective, being correctly placed involves understanding spiritually one's relation to the Earth as a system of which one is an integral part.

The common feature is that, until correctly placed by the values of the particular viewpoint, a proper appreciation of the Earth cannot be achieved. In Box 4.6, the viewpoints differ, they even conflict. Prins is optimistic about technology; Devall & Sessions are optimistic about activism. Murray affirms the Earth through biology, Johnson through philosophy. But this appreciation from all viewpoints involves respect, recognition of significance and, more, reverence. The ability to claim a position of reverent objectivity gives the text an authenticity and lends it, to use another theological phrase, credence.

Other texts contest the ethos of reverent objectivity; they are sceptical of the combination:

> All public movements of thought quickly produce a language that works as a code, useless to the extent that it is abstract . . . The same is true of the environment movement. The favorite adjective of this movement now seems to be "planetary". This word is used, properly enough, to refer to the interdependence of places. . . But the word "planetary" also refers to an abstract anxiety or an abstract passion that is desperate and useless to the extent that it is abstract. (Berry 1990: 197)

For Berry, being objective is a delusion, and it is the reverse of being reverent. But, as his protest implies, the ethos is powerful, it requires and invites rebuttal. The texts suggest Berry underestimates its complexity, its richness and also its availability to different perspectives.

BOX 4.6   Ethos

Theme: reverent objectivity   Topic: Earth

The planet as tragic heroine   "Of all the planets we have colonized totally or in part this is the richest. Specifically: with the greatest potential for variety and range and profusion_{MID} of its forms of life. This has always been so, throughout the very many changes it has – the accurate word, we are afraid – suffered." (Lessing 1979: 14)

The planet as state of being   "Morally we ought, as best we can, to allow the living world, and the entities thereof, in their diversity, to thrive in richness, harmony and balance . . . Thereby we may better live deep and worthwhile lives in a deep and valuable world."_{NCD} (Johnson 1991: 288)

The planet as endless inventor   "The task of recording all of the Earth's life-forms would take the entire working lives of 25,000 taxonomists."_{MID} (Schücking & Anderson 1991: 15)

The planet as fragile genius   "The Earth's biosphere, a thin interlocking layer of land surfaces, oceans and atmosphere, embraces a variety of living organisms so great that most have not yet been named."_{MID} (Murray 1993: 66)

The planet as home   "Many people are beginning to witness and affirm a sane society of humans in balance with the Earth."_{NCD} (Devall & Sessions 1985: 198)

The planet as ultimate system   ". . . Information technology gives us the power to store and process orders of magnitude more material than we could before, and thus the classic product of linear thinking gives us the ability to manage large, multivariate analyses_{NCD} which can begin to model the circular, interactive reality of the planet." (Prins 1993: 185)

Emotive objectivity: population

In population arguments, a key role of ethos is in gaining expert status, but it also involves redefining expertise. The population expert has to present facts, figures, predictions. But in this context, the expert is also presented as engaged, involved, human. Population experts want to have an impact. They aim to energize, to dislodge assumptions, to impose solutions. The result is an ethos of emotive objectivity. There is objectivity in the sense that here are the facts, the figures, the mechanics. But the objectivity is emotive, inherently emotive: the facts are disturbing, *in themselves*, the trends are alarming *in themselves*. The emotion is not personal; it is the facts that are shown to provoke the feeling. If you understand the facts about population, you feel the emotion: panic, terror, anger, urgency.

But normally we assume that emotion is different from objectivity, being objective means not letting feelings interfere. Population arguments are objective, but not conventionally objective; in order to understand the facts *objectively*, we must also feel the response, implies the ethos of emotive objectivity. The facts are the means through which emotion is expressed. The emotions determine the facts, making them relevant to the arguments (Putnam 1981, 1987, Miller 1994).

The ethos can be promoted explicitly, although ethos is most convincing when embodied:

> We can no longer afford to believe that learning the facts or that an emotional response are sufficient unto themselves. Knowledge and emotion have got to be coupled with insight and understanding. (HRH Duke of Edinburgh 1991: 24)

The ethos of population arguments therefore shows something central to environmental arguing as a whole: why facts do not settle the dispute. The facts are inseparable from the emotions; they are emotive facts. And different emotions focus the facts in different ways. In the case of population issues, the emotiveness is related to the nature of the processes underlying the bald population figures and numbers, processes that are both about the cultural centrality of reproduction and the personal nature of the reproductive acts of sex and birth, their relation to the concept of gender and the connection

---

BOX 4.7   Ethos

Theme: emotive objectivity   Topic: population

Pessimism   "So far all human societies have crashed, and ours may eventually suffer the same fate . . . There are four obvious effects of the industrial revolution. The first has been a vertiginous <u>increase in human numbers:</u>_{NID} from around two billion people in 1930, the year of my birth, to well over five billion in 1990. Short of some major disaster, the population will have risen again to over eight billion in the next 40 years. Secondly, and linked with human population increase, there has been a steady degradation of the land surface of the Earth . . ." (Tickell 1992: 95)

Alarm   "I shall probably be accused of being alarmist. While I realize that the scientific community is not unanimous as to the precise magnitude or imminence of the ecological ills portended by exploding population, there is broad consensus that at the very least the <u>prospect of such ills</u>_{NID} is not to be dismissed as nugatory. To do nothing in the hope of some technological miracle would be to court disaster." (Andelson 1991: 29–30)

Despair   "No technical solution can rescue us from the misery of overpopulation. <u>Freedom to breed</u>_{NPD} will bring ruin to all." (Hardin (1968)1989: 299)

Poised ambivalence   "Major forecasts of demographic trends in the twenty-first century are in reassuringly close agreement: <u>growth rates will decline</u>_{NID} steadily throughout the century, leading to a nearly stationary world population of 10 to 11 billion people, or roughly twice the current number . . . This scenario may be comforting or appalling, depending on one's views of the economic and environmental consequences of population size and growth." (Lee 1991: 44)

---

with fears and anticipations of death. Some commentators connect the emotiveness to the nature of the environmental damage caused. For Lovelock, "The problem of feeding a world population of 8,000 million without seriously damaging Gaia would seem to be more urgent than that of industrial pollution" (1979: 114). But for most texts, the nature of population processes is emotive enough in itself without invoking the personality of Gaia. Box 4.7 illustrates this unique flavour, heightened by feeling and flooded by facts.

The Tickell quote in Box 4.7 is exemplary. There is factual material

– numbers, trends, lists – but simultaneously the terms are emotive: societies have "crashed", numbers are "vertiginous". To take an opposing view, Lee's position is expressed in technical demography, the master discipline of population. He again refers to trends, numbers and forecasts, and establishes a consensus. But he also acknowledges an emotional impact; comforting or appalling, there is no way this issue can be neutral. Hence, the professional reinforces his status by recognizing that it is also an emotive issue, even though this time the argument is against assuming a crisis, in favour of reflection on the ambiguities of the figures.

The ethos of population arguments is closely linked to new information discourse: what are the recent facts, the latest forecasts, the trends and tendencies? In so far as new information discourse used in population debates tends to privilege one particular view about population, that of the definition of an overpopulation "problem", so the ethos of emotive objectivity can also reinforce this viewpoint. It need not; one can use informed emotion to support many different viewpoints – a radical ecofeminist approach to population as well that of the population establishment. But the characteristic examples of this ethos, as illustrated in Box 4.7, do link the dominance of the prevailing population viewpoint with this mode of argumentation.

This clearly does not mean that there are no oppositions in population arguments drawing on the new information discourse. Facts contribute to the status of the writer; they don't reduce polarization between different writers and competing texts. And there are deeper causes of polarization, opposing world-views. Emotive objectivity is an appeal adopted in the context of this profound polarization, as information comes flooding in.

Restrained objectivity and its rivals: biodiversity and rainforest

Objectivity and emotion relate differently here. The biodiversity arguments also start with information, information provided by science. That the arguments present information is itself a source of ethos (Putnam 1981, Miller 1994). The information presented is rich, and, as with population, it has two effects: to describe the world *and* to characterize the arguer. The ethos develops differently, though, to imply an arguer motivated by knowledge and the latest data, and to suggest that emotion, although possible, is not the motive. The objectivity is constructed, creatively constructed: such objectivity is an argument, an argument for trust. In the first Grant quote in Box 4.8, key phrases are: "up to a third of" and "the future of humanity". The numeric claim is careful; the reference to "The future of humanity" is impartial, objective. There is only a *trace* of emotion: "sobers the mind". Murray is quantitative, he estimates, he traces trends: "threat" does not much disturb the careful voice. The future is bleak – "If this rate continues" – but the manner is clinical. Shucking & Anderson are intense: "global crisis . . . Catastrophe", but the detail builds up, "15 per cent . . . Ten hectare patch". The manner is precise, not dramatic.

The material could be emotive, we are talking about the loss of the Earth's riches; instead, this time the effect is restrained. The keynote is science, the categories are scientific and they are applied methodically with a hint at deeper tones. The context is enlightenment, a culture based on knowledge. Tensions show in phrases, and uncertainty also emerges: "up to", "random fluctuations", "If this rate continues", "may well support". But in the context of biodiversity, contemporary science is about uncertain worlds, worlds that vary; the data invokes tendencies, not absolutes. The voices are professional, not personal. There is a discourse ethic to biodiversity: being objective is required by the subject. Therefore, objectivity is a value, a positive value, rather than the absence of value judgement (Putnam 1981): by being objective, we overcome the disaster, we find the remedy, and understanding replaces chaos.

Will environmentalism redeem enlightenment? If so, then biodiversity ethos is central. The concern with biodiversity springs from the old enlightenment, from the theory of evolution; there is new information, and new practices, but this is renewing the old languages. At the heart is the objective ethos. Will objectivity be the discourse ethic? There is a closely related alternative that provides a more emotive ethos – more like a population ethos. When the focus is specific areas or species, a different form of discussion results. This can be seen in the media discussion of rainforests.

Rainforests are the exemplary popular image of biodiversity: "The hyperdiversity of continental rain forests is legendary" (Ehrlich &

BOX 4.8   Ethos

Theme: restrained objectivity modified by emotion
Topic: biodiversity

Quantitative objectivity and anxiety   "Why worry about biodiversity losses? The consequences of the <u>loss of up to a third</u>$_{MID}$ of all living species within a stretched generation sobers the mind. This is a <u>scientific, ethical, political and economic issue</u>$_{NCD}$ that is of profound significance to the future of humanity." (Grant 1995: 67)

Explanatory objectivity and gloom   "Large-scale habitat destruction <u>will lead to</u>$_{MID}$ local elimination of plant species and of animals which are dependent upon those plants or the habitat type. Even if small reserves of the habitat are left undisturbed, some loss of species diversity will occur – the small area can support only a relatively small population, at least of larger species, and these smaller populations <u>are more vulnerable</u>$_{MID}$ to extinction as a result of random fluctuations." (ibid.: 77)

Historical and predictive objectivity and alarm   "The greatest immediate threat to terrestrial biodiversity is the loss and fragmentation of natural habitats, especially in the tropics. Major alteration of tropical forests began in about 1600 AD and has accelerated during the past 50 years. <u>In 1980, 113,000 km^2 of tropical forest was cleared but by 1990 annual clearance was 169,000 km^2.</u>$_{MID}$ If this rate continues to accelerate the last area of tropical forest will disappear in about 2040 AD." (Murray 1993: 72)

Objective scrutiny and urgency   "The loss of biological diversity is <u>a global crisis</u>$_{NCD}$ . . . The loss of 15 per cent of the species of the tropical rainforests, however, is a completely different order of biological catastrophe than a similar level of extinctions elsewhere. A ten hectare patch of rainforest in South-East Asia may well support <u>more tree species than</u>$_{MID}$ the whole of North America." (Schücking & Anderson 1991: 22)

Wilson 1991: 759). Rainforest references multiply when biodiversity is discussed and the rainforest incites other arguments, arguments where the ethos is different, and more heated:

Forests have been the green frontier of mankind since time immemorial – a seemingly limitless resource for human expansion . . . But increasingly we are realizing that they offer more than that. They offer more land when we need it; they balance and manage our water supplies and the flow of our rivers . . . (Independent Commission . . . 1986: 17)

The place we call "rainforest" used to be pioneer country, jungle, adventure land, and it used to symbolize origins, timeless and boundless:

The day was beautiful, and the number of trees which were in full flower perfumed the air; yet even this could hardly dissipate the effect of the gloomy dampness of the forest. Moreover, the many dead trunks that stand like skeletons, never fail to give these primeval woods a character of solemnity, absent in those countries long civilized. (Darwin 1959(1845): 280)

Going up the river was like travelling back to the earliest beginnings of the world, when vegetation rioted on Earth and the big trees were kings. An empty stream, a great silence, an impenetrable forest. (Conrad 1974: 48)

Rainforest argumentation now reverses the symbols: the "jungle" was timeless; the rainforest is mortal, almost moribund; the jungle was frightening; the rainforest is vulnerable. In jungle writing, the ethos is awe-struck, slightly appalled; in rainforest argumentation, the ethos is angry, angry at the loss. The anger is partly controlled since the ethos is also scientific: it is Darwin's heirs who are protesting.

Rainforest ethos is more intense than other biodiversity ethos (Box 4.9). The words are angry: "was being incinerated", "was being reduced", "logged, burnt, grazed", "needlessly destructive". There is pathos too, as the rainforest ethos mourns the jungle: "untouched forest", "a vast region of old-growth cedar and hemlock". But the anger is scientific, we are losing the "storehouse of plants and animals", "resilience", "diversity", "old-growth cedar", "ecosystems".

This angry ethos breeds a counter-ethos. The counter-ethos is knowing about the contemporary world and its ways. It is worldly: we know how people are, how the world goes. And it seeks to persuade us: do not fall for concern, do not denounce the modern world in which you live. The North quote in Box 4.10 is ironic, debunking:

BOX 4.9   Ethos

Theme: angry objectivity   Topic: rainforest

The outraged observer   "The landscape changed from *cerrado* to tropical forest. But the forest was on fire . . . All that forest down there belonged to cattle ranchers now, and the new owners had celebrated yesterday's Independence Day by setting patches of it alight . . . The Earth's greatest storehouse$_{\text{MID}}$ of plants and animals was being incinerated as though it was no more than a pile of dry logs." (Girardet 1992: 80–81)

The outraged observer   "There was smouldering forest everywhere . . . And all of this devastation had been wreaked in the name of development$_{\text{MOD}}$ . . . The last great untouched forest region$_{\text{MID}}$ on Earth was being reduced to a giant cattle ranch." (ibid.: 82–3)

Biological protest   "As forest areas are logged, burnt, grazed, farmed and fallowed they may still be classed as forest; but in many cases$_{\text{MID}}$ their resilience is steadily being decreased, their soils eroded and impoverished, the diversity of their natural life curtailed." (Independent Commission . . . 1986: 16)

Melancholy science   "Theoretically the board will have power to curtail$_{\text{MPD}}$ or deny logging permits in a vast region of old-growth cedar and hemlock that has been described by environmentalists as one of the last remaining rainforests$_{\text{MID}}$ in the world, and one of the most threatened$_{\text{MID}}$ by unchecked logging. Chief Francis Frank of the Tla-o-qui-aht tribe said the accord could mean an end to logging methods$_{\text{MPD}}$ that many scientists charge are needlessly destructive of forest ecosystems." (*Washington Post*, 19 December 1993)

---

it diagnoses the rainforest ethos as a psychological hang-up. The key is self-criticism, self-knowledge: recognize your own motives. The chocolate lament is typical: in the end, it's only chocolate, whims and indulgences. But the rainforest ethos also persuades us to face facts, not look away, share the protest. The rainforest ethos is scientific, outraged science; the counter-ethos is knowing, self-aware, mature.

The story of "Rainforest Crunch" provides an example from the media. The proposition is simple: food from the rainforest, is made into gifts and ice-creams, and is sold to help protect those rainforests.

BOX 4.10   Ethos

Theme: worldliness (counter-ethos to angry objectivity)
Topic: rainforests

Psychological self-knowledge   "A sense of paradise-envy~NCD~ accounts for the strength of feeling we have about the rainforest, about pristine environments, for Antarctica, and for the shards of naturalness that remain in our synthetic landscape nearer home." (North 1995: 193)

Moral self-knowledge   "The world stirs to the summoning of its environmental conscience. We berate the fellers of Rainforest, the despoilers of polar wilderness, the whale-slayers, the puncturers of the ozone layer. But who should heed us, we who preside over the near extinction of our pine forest, we who poison our own eagles, we who strangle and devastate~MD~ our mountain terrain?" (*Independent*, 29 December 1993)

Cultural self-knowledge   "More alarming is the mental stagnation induced by two hours of trite pop phrases. Love. Hurts. Lonely. Rearrange as appropriate to form rock lyrics. If you're at a Sting concert, add 'Brazilian' and 'Rainforest' to the choices." (*Scotsman*, 21 December 1993)

Keeping it in proportion   "Chocolate supplies are in danger~MD~ of melting away as the cocoa tree becomes more scarce . . . Experts hope to catalogue~NPD~ what remains of the plant. But they fear its gene pool may be lost~MD~ forever through disease and the felling of the Amazon forest . . . And chocaholics could see their favourite food disappear for good." (*Today*, 29 December 1993)

Beginning Saturday, WTTW, the major public broadcasting service station in Chicago, will become the first PBS station to offer a home-shopping service, peddling merchandise that might be found in a museum gift shop or a fancy import store . . . Like most of its programming fare, WTTW's merchandise will come from Chicago's nonprofit cultural institutions . . . Bronze reproductions of famous art institute lions . . . Or a Rainforest herb-and-food gift basket . . . (Dow Jones, 15 October 1993)

From rainforest argumentation, the counter-ethos rises to the bait:

CC – Commercial Correctness – is peaking in the holiday mar-
ketplace, the year's most intense opportunity for profits . . .
Included are Rainforest Crunch popcorn and Brazil nut bath
beads . . . The surge in socially responsible selling and buying
elevates self-indulgence to self- and endangered-species-indul-
gence . . . (*Washington Post*, 14 December 1993)

The persona is worldly-wise, psychological and satirical. Don't fall
for it, or, if you do, then don't be an innocent, have some irony too.
  The counter-ethos is ironic; reference to rainforests symbolizes
insincerity:

The very same year that the selfless Vermontian ice cream makers
Ben & Jerry introduced an upscale ice cream whose profits
would be used to help save the rain forest, Mother Nature retal-
iated by unleashing Hurricane Gilbert on poor people across the
Caribbean. (*Washington Post*, 14 November 1993)

and there is always the need to make profits:

"There is too much focus on the political aspect of the cooper-
ative and not enough on making it a business," David Alexander
of Cultural Survival, based in Cambridge, Mass., said by tele-
phone. The group has been active in helping the rubber tappers
market their products, chiefly nuts for Ben & Jerry's Rainforest
Crunch ice cream. (*Washington Post*, 25 November 1993)

Meanwhile, angry science considers the value of such exercises in
green entrepreneurship:

Of all the planet-friendly advertising slogans that have emerged
over the last few years, none has been more evocative than "rainforest harvest". The concept links the exotic luxuriance and
impeccable environmental credentials of tropical forests with
ancient and reassuring associations of fertility and abundance.
(Corry 1993: 148)

The principal harvest product – the one which has received most
of the press attention – is Brazil nuts. The flagship of the harvest
. . . Is "Rainforest Crunch", a candy bar containing Brazil nuts

from the Brazilian tropical forest and many other ingredients which have nothing to do with rainforests. Even now, four years after the scheme started, none of the product comes from indigenous peoples. (ibid.: 150)

So, angry science speaks against the product: the ingredients are doubtful, misleading. How does one evaluate anti-rainforest satire and angry science? The arguments are incommensurable, they cannot be weighed side-by-side: it is a cacophony. One side presents images where the Earth is consumed; the other side looks out of the window, at the modern world of consumption.

The role of environmental ethos

Ethos is probably the central factor in environmental arguments. For instance, ethos links truth claims and value claims. Indeed, ethos is the meeting point, the human centre, the rhetorical heart of environmental argument. Truth is not claimed in a vacuum; truth is claimed from a point of view, by an arguer, with an aura of goodwill or anger, calm, or vision, technicality or a common touch. The ethos of truth claims is fundamental in contemporary cultural discussion, precisely because we are confronted by so many truth claims, from different sources, in varying contexts. Often we cannot verify the facts; and we cannot wait for others to do so. Should we then give up trying? Instead we look at *how* the truth is claimed, on whose behalf, with what methods, assumptions, perspective. That doesn't end the argument. If counter-claims arise, then we can assess them too. Facts do count, but in the context of ethos, not outside it. Indeed, strong facts create a strong ethos, and shaky facts are likely to create a weak ethos or an exaggerated ethos. Ethos connects facts and values, because both facts and values are human compositions.

Ethos responds to science and to modernity. An ethos of systematic far-sightedness is one response to science; an ethos of urgent objectivity is another response to both science and modernity – we have the answers, now we must act on them. Objectivity is an ethos, an argumentative persona – or rather many argumentative personae. There is objectivity without emotion; there is objectivity that is also angry or sad at what is seen, although still seeing it objectively; there is objectivity that restrains an underlying emotion. None of this means that objectivity is faked; it means that objectivity is argued, it is represented, as well as performed. The objectivity that counts lost species is different from the objectivity that analyzes projected climate statistics; the objectivity that presents population trends is different from the objectivity that measures contamination, and both differ from the objectivity of a proposal on cost effective pollution control. Objectivity does not mean being value-free; it is being committed, but committed to knowledge, to the clear look at the world.

To analyze environmental arguments is to begin understanding *how* knowledge matters in our discussions and disagreements. Knowledge matters in opposing ways: knowledge matters because it strengthens feelings, and also because it restrains them; or because it makes action possible, and also because it extends reflection. One ethos is urgent, as often on population; another is restrained, although intense, as often on biodiversity. One ethos is authentic, personal witness, as in many rainforest warnings; another is systematic, professional, as in arguments about climate change.

Ethos is not an alternative to validity. Ethos is part of the process of making an idea available to be validated or criticized. Ethos is also a constraint. Arguers need to fit in, they cannot simply start arguing as they would like. We are attuned to some pitches, we cannot respond to others. Acceptability, and conventionality are also involved. There are dangers as well as opportunities. A narrowing of ethos is a narrowing in the possibilities of truth itself.

CHAPTER FIVE

Constitutive figures: metaphorical argumentation, environmental irony, associative argumentation

In this chapter we examine the ways in which certain figures of argument can be used to strengthen and enrich environmental discussions. We focus on three particular figures: metaphorical argumentation, environmental irony, and associative argumentation. And we term these "constitutive" figures because they are creative in use, they generate word pictures, atmospheres, linkages, and they constitute particular shapes of argument.

The use of metaphor can be overt: a direct translation of one thing into another. But many metaphors are implicit, secret, indirect. Metaphors do not merely decorate thought, they actively shape it (Lakoff & Johnson 1980). And metaphors are argumentative: if the Earth is an organism, it is not a spaceship. So, metaphorical argumentation can be negative as well as assertive. Often, metaphorical argumentation introduces world-views: what kind of world is it, how does the world feel? These questions resonate throughout environmental agendas.

Irony is another strategy. Most basically, irony is oblique judgement, particularly hostile judgement. Irony is also about shock, surprise, reversal: progress leading to decline; triumph dissolving in disaster. Irony adapts to subjects and places, and in doing so irony takes on flavours – dramatic irony, trench humour, tragic irony.

Environmental irony has its own atmosphere: it is often heated or intense, it is relatively serious, although capable of cynical turns.

Associative argumentation connects issues, and creates overlapping contexts. By creating these linkages it becomes possible for arguments to move between reference points. The result is a more fluid texture to arguments and a more dynamic process of arguing. Association also has strong psychological roots. It is a key form of poetic thought. Therefore, argument by association can have a very strong emotive pull.

Metaphorical argumentation

Often, it is assumed that metaphors are peripheral, and decorative, that metaphors are outside argument. The assumption is that we think in ideas, and only then adopt metaphors, to manipulate audiences. But metaphors are integral to our thinking and ideas necessarily include metaphors (Lakoff & Johnson 1980, Lakoff & Turner 1989). For example, if you try to think about time, you need to think of flows, drips, grains, slow journeys, quick journeys, circles . . . Furthermore, metaphors affect each other and they interact: if time is a river, then it is not a circle. Often the interaction is argumentative: if one metaphor applies, another falls aside. The use of metaphorical argumentation is well demonstrated in the topics of global warming, population, Earth and resources.

Metaphor making the issue: global warming

In global warming discussions, the use of the metaphor creates and embodies the issue. The central metaphor used is, of course, the greenhouse. On reflection, the greenhouse is a strange metaphor. It implies care, cultivation, concern. Fragile plants are protected by a greenhouse. It is a metaphor in which nature and artifice meet – a greenhouse is a human contrivance to help nature. And the greenhouse effect does refer to a creative process, without which the Earth would never have been warm enough for life. Greenhouses also suggest something fragile, easy to break, but the greenhouse effect is the reverse, a dense and smothering atmosphere. Hence, other metaphors are also used: the thickening blanket, and other metaphors of suffocation.

Throughout, there is a link between metaphor and environmental irony; the metaphor of the greenhouse effect is also simultaneously an environmental irony and it suggests the deadly care, the device to protect, which destroys:

> Without this effect the globe would be much colder – about 33 degrees Celsius colder on average. So, there certainly is a greenhouse effect and it certainly is a good thing since without it the globe would freeze up. The question is whether we want more of a good thing . . . (Read 1994: 33)

The traces of the nurturing greenhouse metaphor linger in the phrase "global *warming*". "Warming" sounds comforting rather than deadly and the tone is relatively restrained, corresponding to the ethos of far-sightedness. So, the central metaphor is compound, carrying multiple associations and suggestions.

As Box 5.1 indicates, there is metaphorical impaction with the play of metaphors, some overt, others implied. The metaphors structure the arguments, they reinforce points, and they make connections. For global warming is about perspectives, horizons, distances and prospects, hanging ambiguously on the edge of the present world. Therefore, there are many metaphorical arguments reconceptualizing time, redefining where the present ends and where the future begins. This use of metaphors follows from the ethos of far-sightedness and the future orientatedness of the issues discussed.

Time and space are indeed unthinkable without metaphors; how can one imagine them except in terms of something else? Global warming focuses the metaphors, and the images of space–time and alters them. The metaphors therefore introduce the question: how to imagine time and space? These are metaphors of global warming, but they are also metaphors of space and time. Global warming encourages the reconceptualizing of space–time, through these argumentative metaphors, from "gaseous greenhouse" to "runaway greenhouse", from "blanket" to rising line (Lakoff & Johnson 1980, Adam 1994, Lloyd 1993).

And there are ambiguities of *care*, for, as we have noted, the greenhouse is a metaphor about care, and warming also suggests care (Heidegger 1993, Gilligan 1982). Whose is the care and why has it been disrupted? This links to the shifting metaphors of the boundary

---

Box 5.1   Figure: metaphors of the issue

Topic: global warming

Greenhouses   "The idea was that the lack of warmth of a <u>cooler Sun could have been offset by a blanket of 'greenhouse' gas</u>_E . . . The trapping of the warmth, which would otherwise escape to space, is the 'greenhouse effect'; so called because it is like, although not the same as, the warming effect of the glass panes of a greenhouse. <u>The first proposal</u>_{MID} that a gaseous greenhouse warmed the Earth was made by a distinguished Swedish chemist, Svante Arrhenius, in the last century." (Lovelock 1988: 73)

"As a result, <u>many studies predict</u>_{MID} that as forests die back, there will be a massive net release of carbon into the atmosphere – 'carbon pulse' – of as much as 225 billion tonnes of carbon . . . The release of such a carbon pulse <u>could lead to a runaway greenhouse effect</u>_E . . . (Jardine 1994: 23)

"By now, of course, the basic mechanism called the greenhouse effect which causes global warming, is well understood. Long before civilization intervened, the thin blanket of gases that surround the Earth was efficiently trapping a tiny portion of the Sun's heat . . . The problem is that <u>civilization is adding many more greenhouse gases</u>_E to the atmosphere and making the 'thin blanket' significantly thicker." (Gore 1993: 89)

"Fewer than a hundred yards from the South Pole, . . . I watched one scientist draw the results of that day's measurements, pushing <u>the end of a steep line still higher on the graph.</u>_{MID}. He told me how easy it is – there at the end of the Earth – to see that this enormous change in the global atmosphere is still picking up speed . . . <u>Global warming is expected to push temperatures up much more rapidly in the polar regions</u>_E . . ." (Gore 1993: 22–3)

---

between human and natural (Merchant 1980, Biehl 1991, Goodison 1994). So, the metaphorical argumentation is used to try and define the indefinable, grasp the contradictory. The metaphors allow complications and questions. They run within the ethos of far-sightedness and the discourses of information, but they also fit with the emphasis on uncertainty, complexity, indeterminacy.

Metaphorical argumentation

Metaphor focusing the agenda: population

Population arguments are heavily visual and the visualization is often metaphorical. Visualization uses the range of images available about the world to present population arguments. By corollary, an image of the world is at stake in population arguments. How is the world imagined? As a small place? A limitless space? A shrinking container, almost filled? How are people imagined? As a crowd? As a swarm? As a deluge? There is also the argument *against* the images, that the images are delusive and reflect only subjective responses, that they have no grip on reality, and that they only obscure the view that they seem to illuminate.

The polarization in population arguments is partly metaphorical between imagining the world as disaster approaches – through metaphors of concentration, shrinkage, swarming – and then un-imagining that world. The disagreement is sharply antithetical, but it cannot be a clear-cut debate. It has tendencies to cacophony, because the metaphors are not really "disputable"; they are presuppositions, composing different worlds, and belonging to different "languages". Therefore, the effect is often scene shifting. The viewpoints are opposite but they do not meet in a debate; rather they represent different imaginary worlds, worlds that are presented and dispelled.

On the other hand, how else can such global issues be represented but through such metaphors? The result is a dispute about representation itself. This dispute is particularly significant where the metaphors are used to support an authoritarian approach to new practices in population policy:

> But the word "population" no longer means people. It typically has connotations of a *bomb* which has released an *explosion* of mainly poor, brown- or yellow-skinned people, who are creating *pressure*, so must be *controlled* – regardless of the methods used. (*The Ecologist* 1993: 141)

As Box 5.2 illustrates, the most vivid metaphors are on the crisis and disaster side of the polarity; on the other side, there are antidotes, criticisms of metaphor, untanglings of metaphor. This approach, metaphorical negation, is well represented in the works of Bookchin:

> If earlier discussions were anchored in rational discourse, the current crop of Malthusians tend to mystify the relationship

---

BOX 5.2   Figure: metaphors of viewpoint

Topic: population

The ascent   ". . . Population growth lies at the core of all trends increasing net environmental stress. It took 130 years for world population to grow from <u>one billion to two billion</u>;_{MID} it will take <u>just this decade</u>_E to climb from five billion to six billion." (Tuchman Matthews 1993: 28)

The wave   "The old global international system, which was formed after the Second World War, <u>is melting down</u>._E I think that this is partly because it was unable to prevent or solve many of the interlinked problems which create environmental insecurity. These are, in my opinion, primarily the consequence of abundant consumption in some countries and deep poverty in many others; of the demographic 'tsunami' (his note: 'tsunami' is the Japanese name for the huge tidal wave that may be created by a submarine earthquake) of <u>exponentially growing population</u>,_{MID} to use Commander Cousteau's arresting image for it; of the <u>rapid erosion of genetic potential</u>;_{MID} of too much, too rapid and careless exploitation of natural resources; of extensive pollution, and deterioration of the natural and the man-made environment; of the steep decrease in the 'cultural diversity' of humankind and growing tensions and local wars in many different parts of the world." (Vavrousch 1993: 89–90)

The explosion   "Of a wholly different character, however, was the situation which observers <u>discovered</u>_{MID} in the Spanish colonies of Central America. Malthus had assumed that the growth of population would be restrained where land tenure limited the supply of land to those who needed it. Much to his astonishment, however, the population was exploding at rates much the same as those to be found in North America." (Harrison 1991: 57–8)

---

between population and the availability of food. Human beings are often seen as a "cancer" on the biosphere . . . (Bookchin 1994: 38)

So the metaphors are buried within the disaster views, and untangled, extracted, interpreted by the anti-disaster view. Taking the whole picture, the authority offered by the ethos of emotive objectivity clearly reinforces the disaster view and the range of metaphors used. The counter-arguers have their work cut out, culturally.

Philosophical metaphors: Earth

Metaphors are fundamental to all self-definition and "Earth" metaphors are central to contemporary identities. If the Earth is like a spaceship, then the Earth is not like an animal; if the Earth is like a jewel, then it is not like a life-support system. Of course, Earth has a material reality. But this does not replace the metaphors. On the contrary, the materiality supports some metaphors and contradicts others. And the metaphors make some aspects of Earth's materiality more relevant and others less relevant. The information discourses are inseparable from metaphorical argumentation. When the astronauts collected data, including visual data from space, this reinforced many metaphors – of a jewel, a ball, a precious sphere . . . Research on homeostatic systems, such as sulphur cycles, reinforces the Gaian metaphor of Earth as an organism, as against Earth as a mechanism (Lovelock 1988, Margulis & Hinkle 1991).

Box 5.3 shows some examples of "metaphorical argumentation". Again there are the two kinds: metaphorical assertions and metaphorical negations. Metaphorical assertions are the proposals: the Earth is like this; we should think of our world in these terms. Often these metaphorical assertions support allied metaphors or a new one may renew an older image (as in the Gore hologram). Metaphorical negation is critical: Earth is not like this, Earth must not be imagined in this perspective. Many metaphorical negations introduce new assertions: Earth is not like a machine, so it is like a body (as in de la Court). Some metaphors are explicit: let me show you an analogy (Gore). Other metaphors are implicit, presented as statements (de la Court). Some examples are both; in the Lovelock quote on Gaia, the metaphor is explicit, and the statement is also literal: Earth is both *like* an individual organism (metaphor) and it *is* an organism (statement).

Lovelock's argumentation is scene-shifting, transfiguring the mainstream and levering consciousness into a new frame of reference. Existentialism refers to "being-in-the-world", the sensation of "here", "now", "who", a sensation preceding all experience, all reflection: Lovelock addresses "being-in the-world" yet does not write metaphysically. Rather, he is specific, citing data. The metaphorical argumentation is therefore connective, connecting facts to metaphysics, data to image. Lovelock represents a style of thought that mixes vision and theory, through metaphorical argumentation.

BOX 5.3   Figure: philosophical metaphors

Topic: Earth variations

Metaphorical assertion: the whole Earth is an individual, and life forms are its cells: "Gaia_{MCD} as the largest manifestation, of life differs from other living organisms of Earth in the way that you or I differ from our population of living cells." (Lovelock 1988: 41)

Metaphorical assertion: the Earth as organism "In the middle of the 20th century, we saw our planet from space, for the first time . . . From space, we can see and study the Earth as an organism whose health depends on the health of all its parts."_{MCD} (WCED 1987: 1)

Metaphorical assertion: the Earth as community (negating Earth as machine) "The world, with all its living beings,_{IE} is not a machine; it is an interdependent community."_{MCD} (de la Court 1990: 108)

Metaphorical assertion: the Earth as hologram (renewing metaphor of life as web): "When the light of a laser beam shines on a holographic plate, the image it carries is made visible in three dimensions as the light reflects off thousands of microscopic lines that make up a distinctive 'resistance pattern'_{MID} woven into the plastic film covering the glass plate . . . Similarly, I believe that the image of the Creator,_{IE} which sometimes seems so faint in the tiny corner of creation each of us beholds, is nonetheless present in its entirety – and present in us as well. If we are made in the image of God, perhaps it is the myriad slight strands from Earth's web of life – woven so distinctly into our essence – that make up the 'resistance pattern'_{MID} that reflects the image of God, faintly." (Gore 1993: 265)

New concepts are advanced, both within specialized fields (here science) and at the level of vision, sensation, and cultural identity.

Altogether the first three quotes in Box 5.3 focus metaphors of the Earth as an organism. These are central to thick Earth viewpoints, where the Earth is dynamic, growing, and changing rather than fixed, a crust, or a shrinking globe. This Earth is alive, interrelated, and dynamic; sick, maybe, but not just denuded or damaged. Implicit in this argument is the idea that the Earth can also heal itself, especially in the quote from Lovelock (new cells, new growth). Although we are also "planetary doctors" (1988: xviii), Lovelock's Earth is self-curing:

> In other words, like it or not, and whatever we may do to the total system, we shall continue to be drawn, albeit unawares, into the Gaian process of regulation. (Lovelock 1979: 128)

The Brundtland extract is also about organic Earth: here the Earth's health needs medical monitoring. And others also stress a human search "to find ways of healing the planet" (Girardet 1992: 16). We have two organic metaphors, with different interpretations. This is a conversational space, with voices playing across the common ground. The "community" metaphor is a variant of this organic Earth metaphor, suggesting a body politic. Here the contrast is between an individualizing metaphor (Lovelock) and a social metaphor (de la Court). But it is also a conversation with much common ground.

These arguments are significant culturally: environmental argumentation is seeking to articulate changes in our sense of being, using the latest theories. It provides reflections on the new world and reflections of the new world. One effect is a polyphony of disciplines – economics, physics, biology, geology – metaphorical argumentation shifting the disciplinary map.

Reflexive metaphors: resources

On resources, the key metaphor is that of the "system". System refers both to knowledge and the objects of knowledge. Nature is a system when viewed by systematic knowledge; systematic knowledge develops once nature is seen as a system. The metaphors link the system of knowledge to the system of nature. Such metaphors are integrated into the arguments. The passages don't say "here is a comparison", as is the case with much metaphorical argumentation about "Earth" or "global warming". Instead, the metaphors are represented as the argument unfolds, and the representation is often diagrammatic as well as verbal. It is easy to overlook the metaphors, because they are so integral.

Limits to growth employs metaphors of accumulation: the rising pile of pollutants, the falling store of resources (Fig. 5.1, Box 5.4). The changes are inverse: as one rises, the other sinks: quantity passes

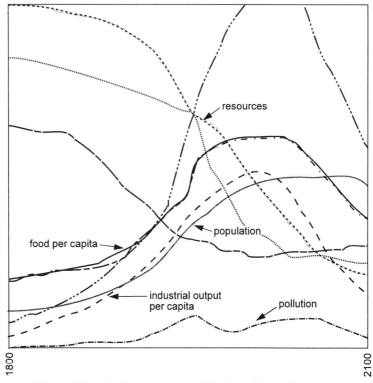

Figure 5.1 *The limits to growth* and visual metaphors of resources.

from the resource pile to the pollutant pile. Eventually, there will be no pile of resources, and a pollution mountain. The graph as used here is a visual image as well as a mathematical representation. In nature, resources and pollution are one system; in knowledge, physical sciences and economics are one system. There is a double integration. The passage argues both that nature must be understood as a unified system; and also that economics must be understood as part of a larger system of sciences.

Daly also integrates economics into the systems of science, using thermodynamics (Fig. 5.2, Box 5.4). Although *Limits to growth* is Newtonian, a world of action and reaction, physical objects and forces, Daly is post-Newtonian, a world of energies and flows. Daly's

BOX 5.4   Figure: reflexive metaphors

Topic: resources

Nature as fixed physical system   "In figure 48 both population and industrial output per capita reach much higher values than in figure 47. As a result pollution builds to a higher level_{NID} and resources are severely depleted, in spite of the resource-saving policies finally introduced."_E (Pestel 1989: 177–8)

Nature as fixed living system   "Standard economics seeks the optimal allocation of resources among alternative uses . . . While not denying the importance of optimal allocation, steady-state economics stresses the importance of another optimum_{NCD} – the optimum scale of total resource use relative to the_E ecosystem. These contrasting visions are represented in Figure 7." (Daly 1992: 180)

Nature as economy   ". . . nature itself provides the archetype of a high-productivity, sustainable economic system (see diagram right). The key to its success is that it is a totally renewable, no-waste economy powered by the sun . . . Humanity in its wisdom has chosen instead a linear economic model in which resources and wastes are disconnected. The inevitable result, as soon as throughput is sufficiently increased, is resource shortages_E . . . (Ekins et al. 1992: 50–51)

Sloping nature: deep nature and accessible nature   "The McKelvey Box diagram (Figure 19.1) illustrates the reserves gradation and the concept of the "ultimately recoverable resource." Feasible natural resources exploitation potential is at its highest in cases that fit the conditions laid down in the top left-hand section of the diagram, i.e. fully demonstrated and measured resource deposits_{NID} that are economic to work. The feasibility of exploitation then deteriorates by stages_E . . ." (Pearce & Turner 1990: 290)

A natural social system   "Many see that it is less a matter_{NCD} of whether a country is centrally planned, mixed (as most are) or capitalistic, since the effects of the underlying industrial model are similar, but more a matter of whether societies are *cybernetically designed* to incorporate feedback at every level of decision-making,_{IE} from the family and the community to the provincial and national levels, from those people *affected* by their decisions." (Henderson 1993: 90)

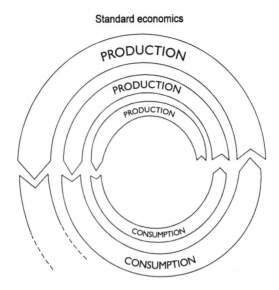

Standard economics considers ever-growing cycles of production and consumption but does not consider the role of the supporting ecosystem. Such a view can encourage an economy that can ultimately strain the surrounding environment.

Steady-state economics

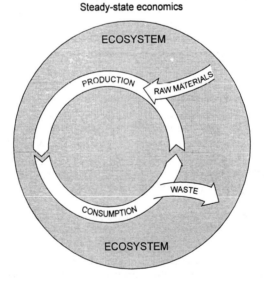

Steady-state economics considers cycles of production and consumption that take the surrounding ecosystem into account and try to achieve a state of equilibrium with it.

Figure 5.2 Herman Daly and visual metaphors of resources.

diagrams represent "two visions", so the metaphors are more overt. The standard vision is pictured in broken circles. The image suggests conflict, collision and even menace. There are spikes, edges. It looks like a broken mechanism. There is also an alarming sense of expansion: the circles get suddenly larger, as well as more jagged. The system is both unstable and closed. Nothing comes in, and yet the parts are in conflict. The alternative vision is harmonious, elegant and symmetrical. The circle is complete. Yet the system is also plural, open to outside elements. Discs and cycles replace the sharp-edged spikes. Daly's diagram is the visual expression of a paradox. The closed system is also fragmentary; the open system is also complete. Daly's vision is aesthetic, as well as scientific; the ecosystem is a circle. The metaphor enables Daly to express a difficult idea, a kind of optimistic Malthusianism. The reservoir is running out, but it is also a beautiful system, and able to run for a long time. Daly's diagram give the case for limits a new feeling, the system becomes consoling as well as constraining. We are in a world that does make sense, it could work, and we could also follow its logic to our advantage, both spiritual and economic.

Ekins's metaphor is also about circles, natural circles and man-made straight lines (Fig. 5.3, Box 5.4). Nature is a circular system, and circles are sacred as well as efficient. The idea of sacred efficiency is powerfully represented, and polarized against human cold inefficiency. The diagrams imply two kinds of rationality, one complex and subtle, the rationality of circles, and the other simple and reductive, the pseudo-rationality of the straight-line world.

Pearce & Turner are stressing the creative potential in resource use (Fig. 5.4, Box 5.4). There are physical conditions, a kind of slope in nature. Down the slope, exploitation is more difficult: "the reserves gradation". But the potential is not fixed by nature; it depends on human capability, how we measure, how we extract and use the resources. Therefore, the potential can actually increase over time, rather than inevitably falling as the pile is used up. Pearce & Turner's gradation is not a product of an iron law, a natural limit; instead, it is the outcome of human science and technology in the context of natural conditions. The diagram is about modern systems and systematic knowledge: "demonstrated", "inferred", "hypothetical"; "geological assurance" not just geological conditions. Even the negative term is hopeful: "undiscovered", not exhausted. The world is

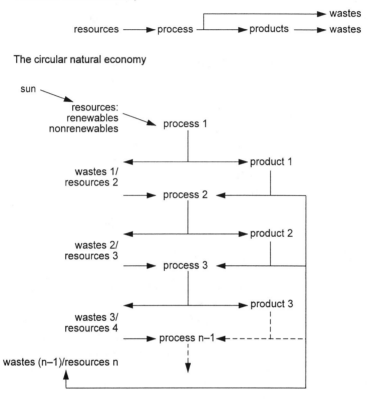

Figure 5.3 Paul Ekins et al. and visual metaphors of resources.

remade by our knowledge of it, and knowledge is not finite.

Henderson is also focusing on human systems. She is concerned with society as a system, an information system. Her text suggests that society is potentially rational, a rational organism capable of processing new information creatively (Box 5.4). The picture tells a different story: the social system is out of control, the complexity is baffling rather than creative (Fig. 5.5). The *Limits to growth* system implied one causal relationship, but Henderson's system is opaque, ambiguous, and elusive; causes are multiple and remedies cannot be simple. The metaphor is entanglement, accelerating disorder, like a

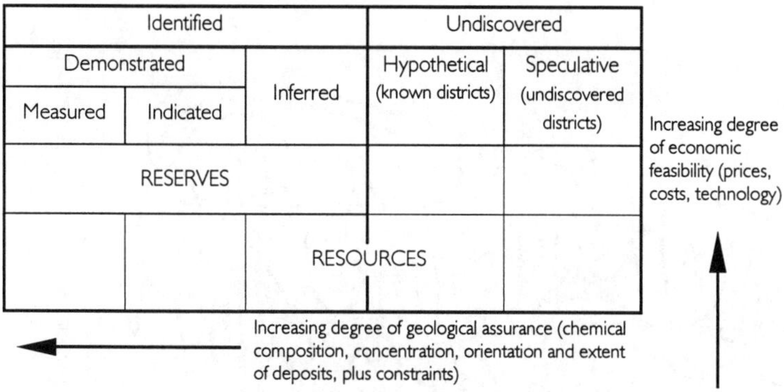

Identified			Undiscovered		
Demonstrated		Inferred	Hypothetical (known districts)	Speculative (undiscovered districts)	
Measured	Indicated				
RESERVES					
		RESOURCES			

Increasing degree of economic feasibility (prices, costs, technology)

Increasing degree of geological assurance (chemical composition, concentration, orientation and extent of deposits, plus constraints)

Figure 5.4 David Pearce and visual metaphors of resources.

gridlocked urban system.

The systems and the metaphors compete: the Newtonian metaphors, the piles rising and sinking; the post-Newtonian metaphors, the circles of energy; metaphors of human systems, of the gradient determined by human knowledge. The systems metaphors give the debate a distinct tone: the sense of increasing complexity, what physical theories call "complexification" (Casti 1995). The viewpoints have become advanced systems, and are self-conscious about being sophisticated. Ultimately, the metaphors are *about* complexity itself, complexity and totality. *Limits to growth* presents complexity that is also comprehensible at a glance; contemporary systems require more glossing. The resources debate asks of systems: are they controllable, do they control us; which is more important, natural systems or human; what is happening to modern systems of knowledge and of social organization? It is these metaphors that make the resources debate so revealing of contemporary culture itself, and of its relationship with the past.

Environmental irony

Environmental ironies show how environmental arguments renew modern culture's engagement with remaking the future. Irony is a

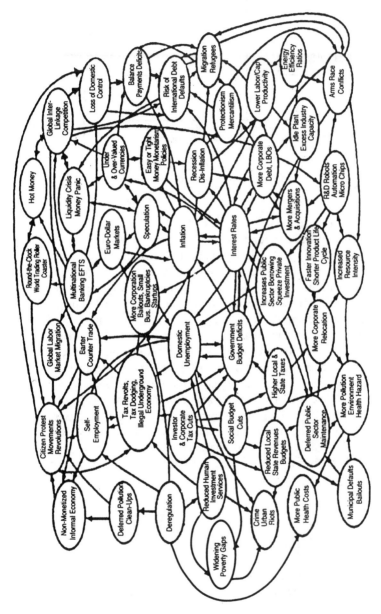

Figure 5.5 Hazel Henderson and visual metaphors of resources: a system view of the global "vicious circle" economy (fast feedback loops).

way to address futures, the futures of modernity itself. Sometimes the ironies are "upside-down" ironies: things that are vital treated as trivial, things that are dangerous treated as secure. Elsewhere, ironies are topsy-turvy: plans make the future riskier; assurance makes the future darker. We explore environmental irony through the topics of global warming, biodiversity, resources and nuclear energy.

Ironies of outcome: global warming

In the global warming discussion, ironies follow from the new timescales involved, especially the redrawing of the map of vast geological time and the sudden link to the present, the everyday. Irony has an inherent connection with time: irony can mean the unexpected outcome – one thing was expected and something else happened. The ironies used in global warming debates are ironies of outcome, the good leads to the bad. In these, the natural effect is made destructive in the future. Such an irony is an appropriate mode for an ethos of far-sightedness – it may look like this, but actually . . .

So the debates concern ironic times and timescales. The past is deep, it stretches back into geological time: suddenly the future replicates the deep past, but speeded up, upside-down. Geological time becomes (nearly) everyday time, particularly in the first two quotes in Box 5.5, from the US scientific report. The implication is that we need to reconceptualize time, to connect the everyday and the geological. The irony is a push towards thinking differently. There is a still deeper irony: warmth turning dangerous. This runs through all the quotes, especially from the US Academy on the "present balmy climate" and Fred Pearce on the warming that cools. The irony that warmth now signals a threat strikes at everyday associations and deep connections. It goes with the ozone ironies: the meaning of a clear cloudless blue sky, and the ambiguities of sunshine. The irony arises from the overturning of associations and values about security and danger. But it is a controlled irony, a medium of education and advice. So, irony is used by far-sighted expertise to unsettle the everyday pictures and values, to foreshadow a new world, but emerging in a gradual way, not as a violent impact.

Critical ironies: biodiversity

There are more fervent ironies. Environmental irony can be an intense argumentative form (Box 5.6). A key practitioner is Vandana

BOX 5.5   Figures: Ironies of outcomes

Topic: global warming

Ironic reversal: the normal creative past is the abnormal destructive future
"From the beginning of Earth's history, certain of these trace gases have had an important role in the principle that is known as the "greenhouse effect" . . . Atmospheric concentrations of carbon dioxide – after water vapour the most plentiful greenhouse gas – have increased significantly, mostly because <u>humans have burned vast quantities of fossil fuels in the past 100 years</u>,$_{MID}$ releasing carbon dioxide in the process . . . If rapid warming of the Earth's surface from changes in the composition of the atmosphere occurs – as <u>scientists warn</u>$_E$ is within the realm of possibility – <u>other global and regional changes could result</u>."$_E$ (NAS 1990: 3)

Ironic reversal: warmth turning dangerous   "As has been the pattern, the cold period lasted about 100,000 years; the present balmy climate is a brief <u>warm spell</u>,$_M$ in a typically icy cycle. This most recent switch from a glacial to a warm phase is of special interest to scientists grappling to make sense of the complexities of modern climate because the amount of carbon dioxide that has accumulated in the atmosphere from when the melting began to the <u>present is roughly equal to the amount</u>$_{MID}$ of greenhouse <u>gases projected to build up from</u>$_E$ the present to about the middle of the next century." (ibid.: 27)

Ironic reversal: from freezing to heating   "It is ironic that, no sooner had we begun to <u>breathe</u>$_M$ easier over the dread prospect of nuclear winter, than we were <u>plunged</u>$_M$ into fearful contemplation of global warming." (Read 1994: 25)

Ironic outcomes: warming becomes cooling   "Global warming could trigger a sudden and sharp <u>cooling across most of the North Atlantic and Europe</u>$_E$ . . . <u>Roughly the difference between</u>$_{MID}$ average temperatures today and at the depths of the last ice age." (Pearce 1994: 20–21)

Ironic outcomes: a beach climate with no beaches left   "The rise in sea level is <u>only one of the consequences</u>$_E$ of a global climatic warming. It is <u>predicted that much of England</u>$_E$ will experience a climate closer to that of southwest France today. The more favourable climate will certainly encourage <u>all forms of coastal recreation.</u>$_{NPD}$ These extra demands will be particularly difficult to meet at a time when decreasing coastal resources will come under increasing pressures from all sides." (ITE/NERC 1989: 21)

BOX 5.6   Figure: irony as critique

Topic: biodiversity

Improvement as destruction   "According to the <u>dominant paradigm</u>_{NCD} of production, diversity goes against productivity, which creates an imperative for uniformity and monocultures. This has generated the paradoxical situation in which modern plant improvement has been based on the <u>destruction of the biodiversity which it uses as raw material</u>."_E (Shiva et al. 1991: 46)

Research as loss of knowledge base   "Centralised research and genetic uniformity go hand in hand in agriculture. <u>It was therefore inevitable</u>_E that CIMMYT (Centre for Wheat and Maize improvement) in Mexico became an <u>instrument</u>_M <u>for the destruction of genetic diversity</u>_E . . ." (ibid.: 47)

Two-faced modernization: we destroy as we are about to advance   "<u>Vast stocks</u>_E of biological diversity <u>are in danger of disappearing</u>_E just as science is learning how to exploit genetic variability through the <u>advances of genetic engineering</u>."_{NPD} (WCED 1987: 148)

Why worry about biodiversity losses?   ". . . As whole populations disappear, another threat looms. This is the removal of the capacity to be resilient, to breed out of danger . . . Just <u>at the time when maximum stress</u>_E is placed on the Earth, so we are losing inherent capabilities of populations to adapt." (Grant 1995: 68)

Shiva. In biodiversity discussions, Shiva is the major ironist, her ironies challenging the professional conversations. But Shiva is herself scientific (she is a physicist); her challenge is professional, but she reinterprets the science.

Shiva stretches the environmental discussion and polarizes it. The first quote argues against applied genetics, the second gives an example. The science of genetics can destroy diversity. The argumentation is ironic: genetics seems progressive, leaping ahead of old practices, but the old practices are more valid. However, the irony is complex, using science against science: Shiva *measures* diversity, and she uses the new measurements to defend old practices. Her science defends tradition, and sees in this a real force for progress. Ironically, Shiva reorientates the professions, and redirects the conversation: do not mistake pseudo-progress for the real thing.

Other texts also have traces of irony. Again, the theme is progress, apparent and actual. Irony heightens the conversation, with little stabs of doubt or anger. Irony is often about bad timing: at the very worst moment, when we need the opposite to happen, *this* goes wrong: when we can use biodiversity, we lose it (Brundtland); when we need biodiversity, it disappears (Grant). The quotes tell ironic stories, ironies of enlightenment: knowledge arrives, but too late; we learn the truth, but cannot act. The question is raised: has modern society ravaged nature before modern science can repair it?

Ironies of appearance and reality: resources

Resource arguments use predictive foresight; they also use prophetic irony: you think you are being prudent, but you are being rash. The ironies hover between tragedy and comedy (Box 5.7). There is a gap between appearance and reality. Destroy your assets, and you seem wealthier; cast caution to the winds, and you appear wise – so run the environmental ironies of Gore and Repetto et al., and *The Economist*. There is also the satirical irony of Henderson: society has been most confident where least entitled, most buoyant as disaster unfolds.

Among the ironies, there is the conflict about economic discourse. Some ironies are part of the argument for reformed economics: Pearce, Repetto, Gore. But Pearce is more radical than Gore or Repetto, more radical and more pessimistic. Henderson is both more scathing and more Utopian; idealism is the other side of her criticism.

Complex irony: nuclear ironies

Nuclear technology has incited a host of environmental ironies. Some ironists argue that the nuclear solution is illusory:

> Given that investment in energy efficiency is at least five times more cost effective as a way of reducing CO_2 emissions, it is truly remarkable that anyone should still be looking to nuclear power as a solution. (Porritt 1993: 34–5)

> Energy efficiency advocates such as Amory Lovins of the Rocky Mountain Institute argue that the cost-effective savings from

BOX 5.7 Figure: ironies of appearance and reality

Topic: resources

The essentials don't count "It [environmentalism] was indeed eclipsed in the United States by Reaganism and in Britain by Thatcherism and by the peak of the <u>dominant paradigms</u>_{NCD} of industrialism, materialism, technological optimism, and general obliviousness to the natural resources on which it was based . . ." (Henderson 1993: 73)

Renewables that aren't "Other things being equal . . . There is reason to suppose that this process of <u>harvesting</u>_M can carry on for very long periods indeed. But, as we shall see, the <u>potential for over-harvesting a renewable resource</u>_E is significant: it is quite easy to make a renewable resource disappear." (Pearce & Turner 1990: 241)

Wastage is valuable "<u>National income accounts</u>_{NPD} (gross national product) take no notice of the value of natural resources: a country that cut down all its trees, sold them as <u>wood chips</u>_M and gambled away the money . . . would appear from its national accounts to have got richer in terms of GNP per person." (*Economist*, cited in Henderson 1993: 75)

Destruction is prudent ". . . <u>The current method</u>_{NPD} for calculating gross national product (GNP) completely excludes <u>any measurement</u>_{NID} for depletion of natural resources . . . a developing nation that clear-cuts its rain forest may add the money from the sale of the lumber to its income, but is not required to <u>place a value on the depreciation</u>_E of its natural resources . . ." (Gore 1993: 337–8)

Squanderers thrive "Even worse, should the proceeds of resource depletion be used to finance current consumption, then the <u>economic path is ultimately unsustainable,</u>_{IE} whatever the national accounts say. If the same farmer used the proceeds of his timber sale to finance a winter vacation, he would be poorer on his return and no longer afford the barn, but national income would only register a gain, not a loss in wealth." (Repetto et al. 1992: 366) energy improvements are potentially vast . . . His customers include US utilities like Central Maine Power, who are allowed by progressive regulatory bodies to make handsome profits saving energy and keeping greenhouse gases out of the atmosphere, while not having to face the headaches associated with building five-billion-dollar coal-fired power-plants (or even more expensive nuclear ones). (Leggett 1993: 56–7)

Other ironists argue that nuclear safety is the more implausible, the more its defenders continue:

Sly was flying over Kalgoorkatta. They dug uranium there, Sly had a lot of money in it. It had been a hell of a battle to get the government to give the go-ahead. People were such a bunch of old women when it came to nuclear energy. It had cost the Earth to prove how safe it was. Dinners, party donations, all expenses paid fact-finding luxury trips. (Elton 1989: 73)

The ironies are often narratives. Tom Bailie has been ill. He has claimed that a local nuclear installation was responsible. Everyone dismissed his views; he was a threat to the local economy. But he turns out to be right:

Today neighbours know that Bailie and all the other so-called lunatics were right. It was the government that had lied. As documents released by the Department of Energy show, during the 1940s and '50s, more than 1,000 civilians were subjects in secret medical experiments involving radioactive substances. At the same time, entire communities, including Tom Bailie's lush valley, were being blanketed by radiation released from the nation's atomic bomb factories. In some cases the emissions were accidental. In other instances the pollution was deliberate. (*Washington Post* , 30 December 1993)

People thought the experts must be believed, but they should have trusted other evidence, they should have questioned the motives. The eccentric was rational; the authority was unbalanced.

Associative argumentation

Argument by association is a process for fixing the terms, the issue, the fact. Association looks outwards as associative arguments place issues, facts, and theories. Other reference points are used to focus the arguments and explain why this one matters. Association is therefore positional. We explore associative argumentation looking at the topics of resources, sustainable development, nuclear power and pollution.

Simple association: nation and resources

In geopolitical disputes, nature meets nationalism. The issues are natural resources and national wealth. Whose resources, which economy? The form is simple: associative argumentation links nation-states and resources.

Resources actually make nations: a nation is entitled to its nature, and association is the means to state the claim. Since the nation-state is modern, resources also constitute modernity. And as resource arguments develop, they connect through the associations with changing arguments about the nation-state. Box 5.8 shows how "small nations" are challenging for a share of nature. It is clear that Estonia needs those mines to count as a nation. The Californian example shows the other side of the coin: to count as a nation, you must be able to share the national resources; otherwise, they aren't national. Southgate & Whittaker span nations: theirs is a regional view. *Irish Times* example shows resources in another sense, not extractable nature but appreciable nature. The other side of all these arguments is universal nature, nature belonging to all or to none, nature as a great system, like Ekins's circles. Associations work their way through the domain of metaphors.

Simple association: historical personalities, current disagreements

One way to make strong personal points, yet appear detached, authoritative or academic, is to express the personal criticism by association with historical terms, to diagnose the other side using historical characterizations. Then the point is more impersonal. And in defending one's own ethos, one can use historical examples positively, again by association. So the arguments are about individual character, yet the form is distanced, impersonal.

171

BOX 5.8   Figure: simple association

Theme: nation and nature   Topic: resources

The oil country   "The SNP is demanding that since Scotland provides the vast bulk of_{NID} the industry's natural resource, justice and logic should dictate that the industry should be led from here, . . ." (*The Herald*, 24 December 1993)

Raw nature, raw nation   "European companies keen to exploit the massive natural resources of Uzbekistan are currently weighing up the president of the country's personal credit_N rating." (*The European*, 3 December 1993)

Minimal nature, shaky nation   "Estonia's only natural resource, the oil-bearing rock, is mined at three open pits, and two underground mines . . ." (*Financial Times*, 2 December 1993)

State of nature, California   "The Clinton administration is about to try to settle_{NPD} another of the nation's longest-running environmental disputes. This one involves the allocation of the most important natural resource in the State of California – water., Northern California is wet." (*The Washington Post*, 13 December 1993)

Natural extravagence   "Latin America is richly endowed with natural resources. Major deposits_{NID} of oil and bauxite, copper and other minerals are scattered from the Rio Grande to the Tierra del Fuégo." (Southgate & Whittaker 1994: 3)

Continuity of nature, continuity of nation   "Our uplands and coasts are an integral part of our national identity . . . an ongoing natural resource of considerable value_{NCD} to the whole country . . ." (Letter from the Mountaineering Council of Ireland, *The Irish Times*, 17 December 1993)

In population disputes, the dominant shadow from the past (and it is clearly a Western or Northern shadow) is undoubtedly Malthus:

> This natural inequality of the two powers of population and of production in the Earth, and that great law of our nature which must constantly keep their effects equal, form the great difficulty that to me appears insurmountable in the way to the perfectibility of society. (Malthus 1970: 72)

Into this complacent world burst Malthus with his assertion that, when population is involved, laissez-faire reproduction does not automatically produce a pleasant world. (Hardin 1991: 164)

In the days of the Industrial Revolution, Thomas Malthus, an unsavoury English parson, formulated his notorious "law of population" . . . It was an unfeeling justification for the inhuman miseries inflicted on the mass of English people . . . (Bookchin 1994: 30)

Of course, it is not really a dispute about Malthus. He is used by association to stand for modern parties in dispute. The associations conflict: it is two languages competing to displace each other. The effect is scene shifting, between one history and another, one characterization and another. Because the details are not really given, there are no grounds for the characterizations, and so it is hard to criticize or revise them. They are imagined worlds, and as such they constitute alternative contexts for the present conflict. Either the context is a history where Malthusians struggle to enlighten a complacent world; or the context is a history where Malthusians meddle on behalf of the powers that be, in their own material interests.

Simple associations: legitimizing sustainable development
Here the argumentative aim is to link sustainable development with other policy issues and goals, and thereby to cement support for the package of goals; one goal lends legitimacy to the others. This can mean sustainable development offering support to more politically difficult goals such as population control, or sustainable development drawing on support from central elements of the policy consensus such as national security. Associative lists of legitimation are well represented in examples drawn from press coverage of US foreign policy statements (Box 5.9).

Looking at the quotes in this Box in more detail, in the first report of a speech by President Clinton, America promises to support the governments of Central America, now that the Cold War is over. But what justifies the pledge, at home in the USA particularly? After democracy, peace and human rights, sustainable development provides a new rationale. Sustainable development also stands for acceptable economic involvement, the reverse of exploitation. In this

---

BOX 5.9   Figure: simple associations

Theme: legitimizing lists   Topic: sustainable development

Sustainable development and . . .   " 'We will not make the mistake_{NPD} of abandoning this region when its dramatic recovery is not yet complete. We will remain engaged_{E} to help Central America attain peace, consolidate democracy, protect human rights and achieve sustainable development,' he [President Clinton] said." (Associated Press, 30 November 1993)

"At a time when AID has had to work with severely reduced budgets, Atwood has sought to redirect the agency's operations_{NPD} toward what he calls 'sustainable development'. He has identified that as a concentration on four basic areas:_{E} population and health, economic growth, environment and democracy." (*Washington Post*, 27 November 1993)

"But Clinton views Asia as the next – and maybe last – great economic frontier_{M} whose growth prospects can surely benefit US exporters. Thus, Joan Spero, under-secretary of state for economic and agricultural affairs, says: 'We aim to promote_{NPD} in the Pacific Rim continued rapid growth, sustainable development and market-oriented economies open to international investment.'" (ibid., 21 November 1993)

---

way, acting in association, sustainable development steals the scene in political argument – steals the scene from the isolationists in America, and from the anti-Americans abroad.

In mediated politics, sustainable development is, therefore, about international relations, the politics of the whole Earth. The administration, recognizing that it needs a new agenda after the Cold War, has reorganized its overseas aid programme to "aim at enhancing 'sustainable development' and 'promoting peace' rather than supporting individual nations." Sustainable development means cooperation and interconnection; it also means economic expansion. So, sustainable development is here acting as a new concept discourse: to fill the conceptual vacuum left by the Cold War and to defend the politics of development, both against anti-aid lobbies and against environmental criticisms of "growth", as well as anti-colonial criticisms. Sustainable development answers simultaneously several different objections to aid, growth and economic intervention: that is why it is such a valued associate for other ideas.

Therefore, sustainable development is a scene-stealer on several fronts in US political argument. The designed impact is that sustainable development sidelines criticism. The effect is twofold: to satisfy the cultural criterion of newness and also to maintain the recognizable viewpoints and policies. This is confirmed in the second quote concerning Atwood, Head of the US Overseas Aid Agency and in the third quote, which focuses on the Pacific Rim economies. In this last quote, imagine the policy statement without the associative linkage of sustainable development. The associated term makes the policy new, as well as recognizable, and also suggests co-operation and concern for the countries themselves: implied is a new idea of the future. It is an avowedly realistic pact of self-interest, but avowedly also ethical, even idealistic, certainly visionary.

Simple associations: delegitimizing nuclear power
As Box 5.10 illustrates, there are two delegitimizing associations of "nuclear": the association with risk and/or accident; and that with war and international politics. Each time the association forms, we have effectively an argument, an argument against extending nuclear power. One crucial point is that every reference to "nuclear safety" implies an associated risk.

Complex association: progress versus regress
A complex associative argument involves the association of pollution with the past and the association of progress with pollution control: pollution is the past, anti-pollution the future.

> The UK pioneered pollution control being probably the first country in the world to introduce systematic controls over air pollution and waste disposal. (UK Government 1994: 37–8)

Implicit in this association is the development of new technological and economic policy measures, which will be even more effective in controlling the problem. Disclosure can therefore always be reacted to and the overwhelmingly reassuring and technocratic ambience can be maintained. The generation of alarm through new information discourses of disclosure can be contained through the operation of

BOX 5.10   Figure: simple associations

Theme: de-legitimizing dangers
Topic: media bite on nuclear energy

Nuclear weapons, disputes   "the communist north's nuclear weapons capability . . . the north is expected to reopen its nuclear facilities . . . an <u>inter-Korea
dialogue'</u>_{NPD} suspended over the nuclear dispute . . . 'We agreed to allow
inspection needed entirely for guaranteeing the continuity of (nuclear) safe-
guards . . .'" (Dow Jones, 31 December 1993)

Nuclear, safety, research   "The European Community commission said it
plans to give the go-ahead to a <u>nuclear safety study venture</u>_{MID} . . ." (ibid., 31
December 1993)

Nuclear, conflict, waste, weapons   "'<u>North Koreans Meet at UN on Nuclear
Dispute</u>'_{NPD} . . . The US wants North Korea to allow the international atomic
energy agency to <u>inspect</u>_{MID} seven sites as well as nuclear waste disposal facilities
. . . The US also wants North Korea to resume <u>denuclearization talks</u>_{NPD} . . .
North Korea wants a pledge from Washington not to use nuclear weapons
against it . . ." (ibid., 29 December 1993)

Nuclear, disaster, tests, accidents   "Nuclear fall-out: About <u>15% of Russia's
territory</u>_{MID} is an environmental disaster area, according to its environment min-
ister Danilov Danilyan . . . Danilyan blamed Soviet-era nuclear tests and accidents at nuclear research centres . . ." (*European*, 31 December 1993)

Nuclear, plutonium, secrecy, warheads   "On December 7 the US Department
of Energy <u>published details of its military stockpile of plutonium</u>_{MID} – revealing it
had used 89,000 kilograms of the nuclear explosive . . . On December 16 the
MoD stated in a written parliamentary reply that 'in the interests of national
security' information about the amount of plutonium used in UK nuclear warheads 'cannot be made public . . .'" (Letter from Dr David Lowry in the *Guardian*, 29 December 1993)

Nuclear, weapons, unknown, small circle   "'US Uranium imported for British
Shells'. The extent of BNFL's role as a weapons manufacturer – hitherto
unknown outside a small circle of nuclear workers and the ministry . . ." (*Guardian*, 24 December 1993)

BOX 5.11   Figure: complex associations

Theme: progress and regress
Topic: the Brundtland report on pollution

Associating air and water pollution with past failings: pollution is the past   "Yet many industrialized and most developing countries carry huge economic burdens from inherited problems$_{NID}$ such as air and water pollution, depletion of ground-water, and the proliferation of toxic chemicals and hazardous wastes." (WCED 1987: 10)

Associating new technology with the future: pollution control is the future   "Pollution control has become a thriving branch of industry in its own right$_E$ in several industrialized countries. High-pollution industries$_E$ such as iron and steel, other metals, chemicals and energy production have often led$_E$ in expanding into the fields of pollution control equipment, detoxification and waste disposal technology,$_{NPD}$ measuring instruments and monitoring systems." (ibid.: 213)

Associating progress and technology against pollution   "Our ecological movement is not against industry, but we must think of the social functions of industries and that pollution and progress are not the same thing. Pollution is not the synonym of progress$_E$ and therefore time has come for new development concepts$_{NCD}$ to come up." (Fabio Feldman, lawyer for victims of Cubatao, WCED Public Hearing, Sao Paulo, 28–29 October 1985, cited in WCED 1987: 215)

---

new practices and the more catastrophic tones, potentially opened up, can be closed again. This is evident in the quotes from the Brundtland Report, where there is a tension between the negative ethos (pollution is alarming, almost overwhelming) and the positive ethos (technology is also impressive, and solutions evolve).

Part of the effect and influence of the Brundtland Report can be explained by the way it proposes cultural redefinition. The argument of the report demands a new culture, new values through the antithetical association of pollution/past against anti-pollution/future. By associative argument, the text positions itself in a debate: many oppose progress and development, and they have used pollution to support their views that progress is dangerous, but we support development, specific development, which addresses pollution problems. To win the debate means to redefine the cultural basis, to relocate progress away from pollution: scene-shifting in defence of new practice discourses of anti-pollution technology.

From constitutive figures to figures of resolution

Metaphorical argumentation links environmental arguments to fundamental themes of space and time, of how the world is imagined and represented. It is responsible for the scope and even grandeur of the environmental agendas. And it is also responsible for the density, the play of meanings, the presence of suggestion as well as precision in environmental arguments. Environmental argument is a key element in our culture's self-analysis about perception and world-construction. It evokes the reappearance of the repressed existential agendas, the agendas of world-making (Giddens 1994: 212).

Environmental irony is about addressing the future, always a major dimension in environmental argument. Indeed, environmental argument is one of the main ways in which our culture seeks to address its possible futures. Meanwhile, associative argumentation makes links across different discussions and debates, bringing broader resonances into environmental topics. On the topic of biodiversity and rainforests, Gore (1993: 244–5) associates science with religion:

> Noah is commanded by God to take into his ark at least two of every living species in order to save them from the Flood – a commandment that might appear in modern form as: Thou shalt preserve biodiversity.

Shiva associates biodiversity with democracy:

> There are, however, not one but many strands in the conservation of biodiversity. The first is inspired by the deep green ethos of the democracy of all life. (Shiva et al. 1991: 43)

Metaphorical argumentation, environmental irony and association together demonstrate how environmental arguments are culturally creative, a creativity that does not depend upon any resolution.

Although not leading directly to resolution, these arguments are immensely inventive, and their inventions are a source of cultural energy and even renewal. The issue of resolvability and resolution is one we pick up and expand on in the next chapter.

CHAPTER SIX

Figures of resolution: dialectic of catastrophe, the feasible possibility, overcoming the polarity

In this chapter we explore the argumentative ways in which resolutions are sought, the ways in which the argument poses solutions. This is not to say that the solutions are always accepted in reality and that resolution of the arguments is achieved. Rather, as we have stressed throughout this book, environmental arguments are deeply irresolvable. But it is in the nature of the arguments that attempts at resolution are made, that solutions are proffered.

Solution rhetoric is created through "figures of thought" (Kennedy 1963, Vickers 1988) and we identify three such figures: dialectic of catastrophe, the feasible possibility, and overcoming the polarity. These figures of thought run between elements of an argument. They are the dynamic connections, locking together ideas, and, as we shall see, each "figure of thought" enacts a logic.

Dialectic of catastrophe

The dialectic of catastrophe is: things are so bad that the solution *must* be at hand, and the solution *must* be absolute. Only a catastrophe justifies the absolute response, the total formula, the necessary plan. The dialectic is that the assumed catastrophe prepares its own antithesis, the solution. The argumentation is twofold: first, there is a catastrophe, imminent and total; then there is a solution, equally imminent and total. The effect is terminal recoil, first the terminal prospect, then the absolute recoil to salvation. There is usually a clear argumentative link to support new practice discourses. We provide analyses of this logic of argument in relation to Earth and population.

Disease into cure: Earth

Dialectic of catastrophe is a logic that interlocks disaster and salvation, "dialectic" referring to the link between opposites that produces a new situation from the conflict:

> No other human generation could ever enjoy a challenge so creative as ours. We have it in our hands to save our one Earth at a time when much of it is on the brink of terminal threat. (Myers 1994: 86)

The crisis is so terrible that the remedy must be equally total, and just as imminent, or . . . At the heart of many environmental arguments, the potent image is of a broken Earth: a planet destroyed, wrecked, wasted. But many arguments also spring back from this point, they recoil from the broken Earth. If the Earth is about to break, we must respond, we cannot wait. Above all, if the danger is total, the salvation must also be total. The dynamic is a negative totality producing a positive totality. The Earth, the world, the planet, are images of that totality. They stand for the whole situation. The broken Earth is the totality of destruction; the response is the sum of all remedies, the totality of restoration. So, texts using "Earth", "world", "planet", project the dialectic of catastrophe. In arguing, they present the totality, the whole crisis and the whole remedy.

Totality sounds abstract. It is elusive and hard to imagine, but the Earth/world/planet variations make the abstract concrete: they give a picture of totality, first negatively, then positively in the dialectic of catastrophe, drawing together new information discourses of disaster and using the momentum to create a positive counter. The more the negative momentum, the stronger it is when reversed – the deeper the disaster, the more compelling the remedy.

Box 6.1 gives examples of arguments linking catastrophe and

BOX 6.1   Figure: dialectic of catastrophe

Theme: disease into cure   Topic: Earth variations

Catastrophe means global disaster; salvation by outstanding people   "Writing this book is part of a personal journey that began more than twenty-five years ago, a journey in search of a <u>true understanding_{NCD}</u> of the global ecological crisis and how it can be resolved. It has led me to travel to the sites of some of the worst ecological catastrophes on the planet and to meet some of the extraordinary men and women throughout the world <u>who are devoting their lives to the growing struggle_{NPD}</u> to save the Earth's environment." (Gore 1993: 1)

Catastrophe means the loss of biosphere; salvation by new systems   ". . . the 1990s will be the decade that will make or break our planet. It will have to be the era of <u>making our farming systems,_{NPD}</u> our fishing and forest policies and the functioning of our cities and factories compatible with <u>the requirements of an intact biosphere_{MID}</u> . . . It is becoming apparent that <u>there is actually very little time left to find solutions_E</u> to the problems of our own making . . . There is still a vast gulf between the <u>Earth's needs_M</u> and what governments are prepared to give." (Girardet 1992: 14–15)

Catastrophe means global pollution; salvation by global political consensus "The onus lies with no one group of nations. Developing countries face <u>the obvious life-threatening challenges_{MID}</u> of desertification, deforestation and pollution, . . . The entire human family of nations would suffer from the disappearance of rain forests in the tropics . . . The members of the <u>World Commission on Environment and Development_{NPD}</u> came from 21 very different nations . . . Despite our widely differing backgrounds and varying national and international responsibilities, we were able to agree to the lines along which change must be drawn. We are <u>unanimous in our conviction that the security, wellbeing and very survival of the planet_E</u> depend on such changes, now." (WCED 1987: 22–3)

Catastrophe means planetary destruction; salvation by end to oppression "<u>Something can be done. The destruction of our planet is_E</u> not based on some <u>mythical law_M</u> of nature but on a conflict, a very human conflict, arising from the domination of some people over other people. But what men and women have done, they can undo or do again differently." (de la Court 1990: 138–9) solution. The logic is Newtonian: action and reaction. But it is also a dialectical logic, leading to a new level. The thesis is catastrophe; the antithesis is solution; the new stage is the future Earth. The sources are diverse: a US Vice-President (Gore), a UN prize-winning environmentalist and programme-maker (Girardet), the Brundtland Report and a counter-report (de la Court). The arguments conflict, even though they share the dialectic of catastrophe. Gore is doom-laden, identifying heroic strugglers, including himself. Girardet is heightened, insistent, and uncompromising about the disaster; the solution is technological. Brundtland is encouraging, practical, responsible, and strategic: the solution is political, with new institutions emerging and a new consensus of nations. De la Court is angry, critical about the catastrophe and defiantly optimistic on the rebound. The dialectics differ, but one can conceive of a debate among these differences. How should the catastrophe be defined? And how should the solution be framed? The questions incite debate. How far do "extraordinary people" take us? How far would global consensus take us? Where do the new technologies fit in? How about conflicts of interest?

Not everyone accepts the logic, that catastrophe creates solution, and total catastrophe creates total solution; the catastrophe can be seen as fabricated, a device:

> One might say that the ozone layer put the Greens on the map. Since other disasters have not yet materialised (there is still standing room on the planet though population levels are still increasing, we are still discovering oil reserves), the discovery of holes in the ozone layer came just in time, . . . (Bramwell 1994: 167)

The argumentation here is "deflating the catastrophe": the planet survives, and provides resources in excess of our demands; the Greens are exaggerating the crisis, to promote their own "remedy". Remove the disaster, and the dialectic is unwound – the proposed remedy appears ideological and dogmatic. Here is another example, from a scientific angle, again "deflating the catastrophe":

> As the human race prepares to venture into a new century, conversations and news reports are peppered with references to our "fragile and endangered planet". This phrase almost certainly

exaggerates the case. The Earth is 5 billion years old, and over
eons it has endured bombardment by meteors, abrupt shifts in
its magnetic fields, dramatic realignment of its land masses . . .
(NAS 1990: 1)

NAS wants action, but resists the dialectic of catastrophe, favouring
precautions and adjustments. This is based on the ethos of reverent
objectivity reverence derived from scientific (and unexaggerated)
knowledge of the planet's history. It uses the resources of new information discourses and circulates around a thick Earth viewpoint.
The resulting "deflating" argumentation is scene shifting, away from
the catastrophists. The premise changes, the terms are transferred
and the register switches. Bramwell and NAS are not contradicting
Gore or Brundtland, they are dismissing them. It is the common
forms that relate the arguments to one another, even though it also
reveals unresolvable gaps and gulfs.

On the cusp: population

The polarizations of this form of argumentation can occur along two
axes: different logics of terminal recoil, different catastrophes and
contrasting salvations; and winding up the dialectic of catastrophe
versus unwinding the dialectic of catastrophe – if there is not a disaster around the next turning, then there isn't an absolute salvation,
there is no master blueprint. These alternative positions can be seen
in population debates in terms of both alternative diagnoses and
alternative remedies, and in terms of reinforcing or defining the topic
of the arguments.

This is a debate between participants. Those who disagree on the
logic of terminal recoil can ask: What should the plan be? Should it
be managerial, or should it be welfare based? Should it be hierarchi-
cal or egalitarian? How far is it redistributive? And there are more
and less authoritarian responses and solutions as apparent from Box
6.2, with Hardin representing the overtly authoritarian position and
the Prince of Wales a more humane presentation. In posing resolutions to the arguments, the contributors come closer to an engagement than in any other aspect of population argumentation. But it is
not an engagement that will readily lead to a resolution of solutions.

BOX 6.2 Figure: dialectic of catastrophe

Theme: on the cusp   Topic: population

Winding up for terminal recoil

RUIN BY ACCELERATION; SALVATION BY PLANNING   "For the great majority of mankind still experiencing, high rates of population growth, action now to initiate, or to accelerate, fertility decline is imperative." (McNamara 1991: 65)

RUIN BY EXPLOSION; SALVATION BY RECOGNITION   "Unless we accept, that the population explosion, is the most serious, predictable and intractable crisis facing us, we shall not be able, to avoid it." (Brundtland 1993: 2).

RUIN BY FREEDOM; SALVATION BY ABSOLUTE CONTROL   "No technical solution can rescue us from the misery of overpopulation. Freedom to breed will bring ruin, to us all . . . The only way we can preserve and nurture, other and more precious freedoms is by relinquishing the freedom to breed, and that very soon." (Hardin 1968 (1983: 299))

RUIN BY (POOR) POPULATION; SALVATION BY MODERN WELFARE SYSTEMS ". . . The factors, which will reduce population growth are, by now, easily identified: a standard of health care that makes family planning viable, increased female literacy, reduced infant mortality and access to clean water." (HRH Prince of Wales 1993: 6)

Unwinding terminal recoil

NO ECONOMIC RUIN BY POPULATION; NO NEED FOR SALVATION FROM POPULATION   "Some people justify population control policies on the grounds that population growth reduces economic wellbeing. It is not clear, however, whether population growth strongly affects such variables as prices, incomes, unemployment, balance of trade . . . In any event, even if population growth were known to reduce the wellbeing of future generations, it would not automatically follow that such growth should be slowed." (Lee 1991: 315)

Overpopulation is subjective; the solutions are subjective   "Some will argue that the world is already overpopulated, that no further growth is required. It is of course every person's right to venture his own opinion of what world population should be, and to campaign as vigorously as he may wish on its behalf. If we are to recommend imposing our ideas on the rest of the world, however, something more is required. On what grounds is it claimed that the world is overpopulated?" (Feeney 1991: 74)

The feasible possibility

The feasible possibility is about what must be done: the key is to connect the necessary, "what must be", to the possible, "what can be", and to review the alternatives against these criteria. A key category in classical rhetoric is the possible and the impossible, especially in deliberative, decision-orientated arguments; "the feasible possibility" is a contemporary variation, associated particularly with new practice discourses on environment. In demonstrating the feasible possibility, the argument expresses the practical question of how to solve the problem. It defines the area of solubility and emphasizes the contest again between different rationalities – from scientific to political, from technological to managerial.

The form of argumentation is as follows. A problem arises, which is serious but the tone is not as in the "dialectic of catastrophe"; there is a sense that scope exists, choices must be made, and it is a practical question. A deliberative stance is adopted: let us review the options; let us see what is possible. But then it emerges that, in the context, there is one option that is essential, one choice that has to be pursued. The field narrows and the feasible possibility is argued.

Applicable rationality:
global warming and nuclear technology

Global warming attracts arguments of "the feasible possibility". The result is less heightened than, for example, population issues, but nevertheless requires solution. So, we have a process of review, a posture of consideration, and then a spring into choice, *the* choice. It is a persuasive thought sequence, a figure of thought like the dialectic of catastrophe, but in a lower key. Sometimes the range isn't directly surveyed, but there's a tone of consideration, thinking it through, and then pointing to the necessary single option.

Often this figure results in a link being made between discourses, and particularly between new information discourse and new practice discourse. There is a feeling of practicality and also of solubility. But there are tensions. The solutions differ and can conflict. There can be a major conflict between rationalities: technological, institutional, or political. So, the global warming arguments focus on the nature of rationality itself and the location of rationality in the world culture. This can be explored through a paradigm example (Bertram

1992: 423–4). First there is the account of the problem and the identification of a number of proposed solutions:

> The past two years have brought a flurry of scientific and diplomatic activity in the quest for an international convention to limit global emissions of the gases which are responsible for global warming.

But there is a gap since the solution has not been formulated:

> While there appears to be widespread agreement on the need for policy initiatives, no consensus has yet emerged regarding the best mechanism through which the international community could implement a programme of cutting back global emissions.

A review of the possibilities follows:

> The three leading contenders are direct quantitative emission restrictions, carbon taxes, and tradeable emission permits, . . .

and the feasible possibility is announced:

> Bertram, Stephens and Wallace (1989) argued that a worldwide tradeable-permits system could be an effective way of advancing the interests of the developing countries in harmony with the global community's interest in protecting the atmosphere . . . This article develops the suggestion . . .

In this instance, it is institutional rationality, the rationality of frameworks, arrangements, negotiations, that is adopted. In Box 6.3 other alternatives are proposed. It is noticeable that scientific rationality is an element in most solutions. The mixing of scientific with other rationality is a powerful theme in feasible possibility argumentation.

In the nuclear examples given in Box 6.4, feasible possibility argumentation shows vividly its three elements: dismissing the alternatives; focusing a single solution; asserting that it is plausible and realistic. Walter Marshall dismisses coal and oil, the main rivals; he then argues both that nuclear technology solves the problem, and

BOX 6.3 Figure: the feasible possibility

Theme: applicable rationality   *Topic: global warming*

The educational solution "Some would argue that the public need only express concern, not understand all the details of global-warming – the scientific, political, and industrial leaders can put together appropriate responses by themselves. I would challenge this argument for two reasons. First, just as citizen concern is needed to motivate leaders, so also is some citizen knowledge needed to keep leaders on the right track. My second apprehension about today's state of public knowledge is this: if world leaders decide to reduce greenhouse emissions by two-thirds, such a large reduction will require consumer and worker cooperation as well as citizen consent that major societal changes are worth the effort." (Kempton 1991: 207)

The institutional solution ". . . the potential cumulative effects$_{MID/E}$ of global climate changes, suggest that it is time$_E$ to develop and/or amend the necessary legal framework,$_{MPD}$ at both a national and a regional level, to assure the protection and preservation of the marine and coastal environment." (Gable et al. 1991: 411–12)

The scientific–managerial solution "One of the more dramatic consequences$_E$ that can be expected from a global warming is a rise in mean sea level$_{MID}$ . . . Deleterious changes could be offset to a considerable degree by appropriate management$_{MPD}$ soundly based on research." (ITE/NERC 1989: 3–4)

The scientific–technological solution "Thus the onset of global warming, and the incentive it provides to use non-fossil alternatives, both brings forward and changes the direction of the expected transition$_E$ due to the exhaustion of the cheaper oil reserves. A renewable energy raw material, i.e. a sustainable fuel supply, is needed, powered by the solar energy that is captured in the photosynthesis involved in growing trees or other biomass." (Read 1994: 82)

The scientific–technological solution "This analysis has shown that the near-term transportation alternatives – fossil fuel-based methanol, natural gas, LPG, corn-based ethanol and current-technology BPEV's using conventional fossil-fuel electricity – generally would not provide large reductions$_{MID}$ in emissions of greenhouse gases . . . On the horizon, however, are alternatives$_E$ that could provide substantial reductions . . . Fuel cells could be a key$_E$ mid-to-long term technology$_{MPD}$ in both electricity and transportation." (DeLuchi 1993: 191)

BOX 6.4   Figure: the feasible possibility

Theme: applicable rationality   Topic: nuclear energy

Plenty or scarcity   ". . . Accessible sources of convenient fossil fuels, like oil and gas, will become rarer$_{MID}$ some time in the first half of the next century. I therefore remain convinced that nuclear power is the ultimate source of energy for mankind.$_E$ Its use will start to grow in the early decades of the next century and will dominate by the year 2050." (Marshall 1993: 147)

The lesser fear   "As burners of fossil fuels begin to be charged and taxed$_{MPO}$ to reduce the pollution they cause, that will militate in favour of alternatives, such as nuclear power, especially if our fear, of radiation is to some extent replaced by our fear of global warming." (North 1995: 81)

The real world   "'Ukraine Turns to Nuclear Power'. Because of a harsh winter and Russia's decision, to raise the prices of oil and natural gas by 25% to the full world level, Ukraine has turned to nuclear power. Parliament voted October 21 to reverse its earlier decision to close the Chernobyl plant . . ." (Associated Press, 31 December 1993)

The safe risk   "The health risks for the development of peaceful nuclear technology, including nuclear electricity, are very small when compared with the benefits from the use of nuclear radiation for medical diagnosis treatment." (Ian Wilson, Vice-President of the Canadian Nuclear Association, in Brundtland: 184)

---

that it is reliable and realistic. The Ukrainian government argues that nature, money and politics make alternatives impossible; they must have nuclear power. It is cheaper, it will, by implication, work. Some steps are left implicit here, but the logic is feasible possibility. The Nuclear Association argument is that nuclear technology works, not only for energy generation but also from a health perspective; it is safe, and it is credible. There is no alternative source of diagnostic treatment as good as radiation. But solar arguments also have the same elements: dismissing the alternatives; proposing the solution; asserting practicality.

The result is therefore a cacophony of vying solutions. They cannot all be the *only* feasible possibility; yet they present themselves as such. Viability and practicality are used to press for a unique path through the rationality of environmental issues. The options available within the cultural world are limited through feasible possibility argumentation.

Pragmatic ideals: pollution

The feasible possibility is repeatedly used in the new practices discourses of environmental economics and pollution, as exemplified by the three quotes from Royal Commission on Environmental Pollution in the Box 6.5. By demonstrating the feasible possibility, stress is placed on both the way in which the method of pollution control is available and possible, and on the overwhelming benefits of this approach in terms of dealing with the necessary problem in a practical way ("essential foundation . . . Will make it possible"). It thus also builds on the new practice discourses, as indeed many of the terms used in that discourse imply (BPM, BPEO, BATNEEC). And it further reinforces the position of the regulator within the debates, a viewpoint supported by the prevailing ethos.

One danger in this argumentative form is the tendency to create straw men, particularly when arguing for market-based instruments. The picture of regulation or command-and-control offered is often a caricature, and the necessary assumptions of the market-based instrument alternatives are often glossed over to promote the environmental economists' chosen options (the first quote shows these tendencies at work). Hence, the offer of reassurance in such policy debates is sometimes based on an overly elegant simplification. The justification here is the pragmatism that such environmental economic "solutions" can offer. They are more realistic than either the techno-fix or the ethical application of the precautionary principle (but for a critique see Jacobs 1994, and also Malueg 1990 and Oates et al. n.d.). The fourth quote represents a critique of environmental economics' new practice discourses and their argumentation, a critique that inverts the feasible possibility and applies it negatively ("never costless . . . Effectively infeasible").

BOX 6.5   Figure: the feasible possibility

Theme: pragmatic ideals   Topic: pollution

The balanced way forwards   "A <u>sustainable transport policy</u>$_{NCD}$ can not, in our view, accommodate growth of more than about 10% a decade in overall demand for transport. At this rate of growth <u>there are two possible forms</u>$_E$ that in theory such a policy could take. The one we are advocating, is a <u>balanced combination</u>$_{NPD}$ of pricing and measure to promote alternatives to private road transport . . . The alternative would be to raise prices (for fuel and/or road use) more sharply than we have recommended, and rely entirely on that to hold down road traffic levels. This would be open to strong objections on grounds of equity. It is also likely to be <u>less effective in achieving</u>$_E$ environmental objectives, especially in urban areas." (RCEP 1994: 252)

The compatible way forwards   "These changes of direction provide the essential foundation for a sustainable transport policy, and <u>will make it possible</u>$_E$ for the economy to develop in ways which . . ." (ibid.: 234)

The affordable way forwards   "Finally, BPEO gives explicit recognition to the need for pollution problems to be <u>tackled in a way</u>$_E$ that is economically <u>practicable</u>$_E$ as well as <u>technically feasible.</u>"$_{NPD}$ (RCEP 1988: 42)

The realistic way forwards   "I noted above that legislators might ignore economists' advice because the analysis on which it is based is not relevant to actual conditions, and I cited particularly in this connection the question of monitoring. The economic analysis of the choice of pollution control <u>instrument typically ignores this question</u>,$_{IE}$ in effect assuming that monitoring is feasible and costless. In fact, <u>the continuous monitoring required</u>$_{NPD}$ for the taxation of emissions (or for regulation of such, or for the operation of a permits market) is never costless and is in some cases so costly as to be effectively infeasible." (Common 1989: 1306)

Note:

RCEP – Royal Commission on Environmental Pollution
BPEO – best practicable environmental option

192

Overcoming the polarity

Overcoming the polarity is central in environmental discussion. The argument works by making a polarity between two factors and then advancing beyond the division: economy and nature as poles to be reunited; looking to tradition and to the future and then moving to a resolution.

Resolution by reconciliation: sustainable development, sustainability
Sustainability and sustainable development imply a need of reconciliation, reconciliation between environment and economy, between the spiritual and the practical, and also between different interests. In each case, the "figure of thought" is division then synthesis, "overcoming the polarity". It is important that this figure of thought both implies the division and enacts the reconciliation.

Because these types of reconciliation are so central to sustainability and sustainable development, the dominant form of argumentation used is concerned to overcome objections to the principle of pursuing sustainable development. This is why the *necessary* form is overcoming the polarity: sustainable development as "a unifying motif" (Shrivastava 1993) or as "an overlapping consensus" (Young 1990) or even "a pre-emptive consensus" (Myerson & Rydin 1995). This is illustrated in the Box 6.6 in the Pearce *Blueprints* and the UK *strategy for sustainable development*. The explicit polarities involved here relate clearly to the cleavages in the sustainable development discussion outlined above between stronger and weaker sustainability, between different viewpoints relating development, environmental protection and equity. The role of sustainable development is to suggest ways across these cleavages:

The phrase "sustainable development" has staying power because most people want to believe in it. It survives because it appears to build bridges between the demands of environmentalists and developers. (Pearce 1993: 183–4)

. . . the Community shall have as its task "to promote throughout the Community a harmonious development of economic activities, sustainable growth which is non-inflationary and respects the environment; (The European Union [Maastricht] Treaty 1992: Article 2)

BOX 6.6   Figure: overcoming the polarity

Theme: resolution by reconciliation
Topic: sustainability and sustainable development

Between social and environmental priorities   "The social dimension states simply, but powerfully,[E] that a sustained society is also a truly democratic society
with rights of expression, dissent, participation, self-reliance and equality[A] of
opportunity. Political and economic structures have to deliver social as well as
environmental sustainability." (Pearce 1993: 185)

Between economic and environmental priorities   "The modern sustainable
development debate[NCD] has tended to shift the focus[E] away from growth *versus*
the environment to one of the potential *complementarity*[E] of growth and environment." (Pearce et al. 1989: 21)

* "Most societies want to achieve[E] economic development to secure higher
   standards of living, now and for future generations.
* They also seek to protect and enhance their environment, now and for
   their children.
'Sustainable development' tries to reconcile these two objectives." (UK Government 1994: 6)

"Sustainable development does not mean having less economic development:
on the contrary, a healthy economy[N] is better able to generate the resources
to meet people's needs, and new investment and environmental improvement
often go hand in hand." (ibid.: 7)

The quotes show clearly how the attempt to resolve conflicts
between goals for these three elements of policy drives the arguer
towards a central position on sustainable development, away from
what are cast as the extremes of the limits to growth or quasi-cornucopian schools. This "synthesizing" makes the sustainable development discussions fairly unified compared with those over many other
"environmental words" although not, it must be stressed, a consensual one. It is based on invoking a "reasonable" version of the essential vision, on emphasizing the moderate and optimistic nature of the
concepts. These ethos-based characteristics of the arguments – reasonableness, optimism, moderation – are one of the means by which
polarities are overcome in sustainability discussions.

Overcoming the polarity

What is involved in these kinds of rhetorical strategy is a scene-stealing move, away from those who posit a conflict, a move that works by denying the inevitability of such a conflict, by arguing that it is possible – conceptually as well as practically – to have environmental sustainability and economic development, environmental sustainability and political democracy or social welfare measures. The quotes from the UK *strategy for sustainable development* in Box 6.6 indicate this scene-stealing strategy. In the first of these quotes, the arguer trumps the opponent who argued that we cannot do it, or that we should concentrate on the present or that development is undesirable: "most societies" treasure both growth and ecology, and sustainable development is the modern expression of this universal will. The phrase "most societies", especially, has an argument-ending quality that is typical of the scene-stealing strategy. In the second quote from the UK government on economics, again the rhetorical strategy involves trumping the opponent who argued that industry is the enemy of environment. It enacts the end of an exchange with an opponent who believes industry is dispensable, wealth is dubious. The key word is "healthy", in "healthy economy"; the scene is stolen, wellbeing is integral to the economy, the spirit advances through economics.

Another potential conflict that must be overcome, and even at times denied by sustainable development arguers, is that between the North and South, the conflict that received so much attention at the 1992 Rio Earth Summit and is the implied target of the Brundtland Report title *Our common future*. Here the unifying theme of sustainable development is seen as essential, if not always fully effective, in progressing international agreement on specific sub-issues: global warming, biodiversity and population levels. Here the globalization of the arguments presents a scene-shifting strategy, as illustrated by the Pearce quotes in Box 6.6. This involves a tendency to try and alter the bases of discussion rather than feed into established assumptions. The extracts in the Box show how a new concept discourse is used to scene-shift, to redefine the context so as to begin a new type of discussion: let us talk differently, given the new situation, the growth of contemporary population, the energy crisis, the new views of culture; let us talk differently because the problems are now global; let us begin a global discourse, a global argument.

And other forms of polarity may also be unified in the sustainable development debate, cultural and even philosophical polarities:

We hope that we have gone some way to demonstrating that "sustainable knowledge" must be a mixture – of the social, the scientific, the local, the technical, the natural, and perhaps even the magical – that refuses *a priori* to privilege science . . . To speak of "sustainable knowledge" is to begin to speak of the local and the general, the natural and the social, Western and non-Western cultures in the same breath. (Murdoch & Clark 1994: 129–30)

Overcoming the polarity is the richest form of argumentation applied to sustainable development and sustainability. The figure of thought is complex, flexible and able to move between contexts. It has done much to locate these environment words at the heart of cultural argumentation.

Beyond trade-offs: pollution

Overcoming the polarity with pollution involves an attempt to argue that it is possible to have the fruits of economic activity while minimizing or containing the pollution arising from it. The language of environmental economics is a key resource here, as Box 6.7 shows, although there is a consciousness exemplified by the last quote from Lawrence Summers, that economics may be in competition with other disciplines and other criteria, including ethical criteria.

The "polluter pays" principle is a key new practice discourse using this argumentative form, exemplified by the first quote from Royal Commission for Environmental Pollution. By connecting scientific and economic measurement, the PPP rearranges the cultural context: science will provide the primary measurement of pollution, then economics will translate that measurement into money values. The impact is transformative: values are made commensurable, so that society translates sulphur concentrations into money values. PPP proposes a science-based market, making the market take account of science. The result is "optimal", a better trade-off between the costs and benefits of economic activity. The dominant ethos underpins the use of PPP to suggest that existing systems can respond, they can accommodate new values.

If economics is linked to ethics in this way, and science to politics, then words move around, relative to one another. For instance, the word "damage" shifts, once we apply the economics of "optimality" to pollution. Ordinarily, "damage" sounds absolute, a bad thing, to

BOX 6.7   Figure: overcoming the polarity

Theme: beyond trade-offs   Topic: pollution

Present costs into future benefits   "The costs of the policies, we recommend, over the next decades of transition to a sustainable system of transport and access, are not large in relation to the seriousness of the environmental problems. But as costs of resolving problems inherited from earlier generations, bringing benefits which will be experienced by generations in the future, they should properly be borne by the community as a whole, out of public expenditure." (RCEP 1994: 254)

Value for money, maximum technical impact   "The European Community's Fourth Environmental Action Programme specifically recognises that a sector by sector approach to problems may not achieve the maximum pollution reduction for a given economic cost." (RCEP 1988: 2)

Business and society   "The CBI has pointed out that investment in good environmental management makes sound business sense, not only by improving a company's reputation but by reducing costs through increased efficiency and the avoidance of waste . . . There are economic as well as social benefits to be gained from such an approach." (ibid.: 3)

Economic logic and ethical logic   "The measurement of the costs of health-impairing pollution depends on the foregone earnings from increased morbidity and mortality. From this point of view a given amount of health-impairing pollution should be done in the country with the lowest cost, which will be the country of the lowest wages. I think the economic logic behind dumping a load of toxic waste in the lowest-wage country is impeccable, and we should face up to that." (Lawrence Summers, internal World Bank memo, 12 December 1991, quoted in Foster 1993: 10)

Note:

CBI – Confederation of British Industry
RCEP – Royal Commission on Environmental Pollution

be avoided. But the word "damage" adjusts, it becomes relative and even "worthwhile":

> ... The costs of controlling acid rain are likely to be less than
> the benefits obtained from reduced pollution damage, although
> more work is needed to verify this provisional conclusion ...
> The cost–benefit approach ... Essentially says that some envi-
> ronmental damage is worthwhile because the loss to society of
> going beyond the "optimal" amount of damage is greater than
> the gain. (Pearce 1993: 61, 62–3)

The idea of optimal damage gains a currency where previously it would have been held as false coin, an oxymoron.

Methodical synthesis; biodiversity

Conflict is also resolved in the name of biodiversity. The conflicts are relatively muted and less emphasis is put on the contradiction than on the overcoming. The effect is, therefore, more consensual. The overcoming is more to be expected, less to be asserted. The key is method, professional and scientific method, rather than sweeping new perceptions or logics.

In Box 6.8, Murray is overcoming specific contradictions: first, the contradiction between economics and the spirit; and secondly, the contradiction between nature and society. In the first quote, he refers to Chief Seattle, to whom is attributed a saying about the holiness of nature. The key word is "impossible": is it not inconceivable to esti-mate holy nature, to estimate nature financially? But there is no alter-native: only monetary value is taken seriously; therefore nature must be counted in order to count. The second quote uses the word "wild" to contrast nature and society, with its "uses"; again, the argument overcomes its own contradiction, since when nature is useful and exploitable, nature is safe from overexploitation. Pearce & Moran (1994) develop the logic that nature must be valued economically. Society needs a new economics, an economics that respects nature: a new practice of environmental economics will be able to overcome the polarity between the social and the natural. The key remains meas-urement: economics will provide the means to measure nature, and the resulting numbers will be persuasive in advancing conservation.

The Grubb and Brundtland quotes concern global politics. Grubb

---

BOX 6.8   Figure: overcoming the polarity

Theme: methodical synthesis   *Topic: biodiversity*

Polarity: intrinsic value and economy; synthesis: new valuation   "There is a growing realisation that, the present variety of living organisms will not be maintained, unless the value of biodiversity is revealed and appraised in terms corresponding directly with those used in the global economy. So, paradoxically, immeasurable, intrinsic value may only be protected in the short term by finding ways to do what Chief Seattle thought_{NCD} impossible: placing extrinsic value on biodiversity." (Murray 1993: 67)

Polarity: economy and conservation; synthesis: economic justification   "Two principal uses_{NPD} of wild plants and animals are cited as providing a strong economic justification for conservation of global biodiversity." (ibid.: 74)

Polarity: developed and developing; synthesis: interdependency   "The interdependency of the developed and the developing nations in maintaining_{N} biological diversity is recognised_{NPD} [by the Biodiversity Convention] . . ." (Grubb et al. 1993: 76)

Polarity: nature and development; synthesis: new policies   "Third World governments can stem the destruction_{E} of tropical forests and other reservoirs of biological diversity while achieving economic goals . . . Reforming forest revenue systems_{NPD} and concession terms could raise_{A} billions of dollars of additional revenues, promote_{A} more efficient, long-term forest resource use, and curtail deforestation." (WCED 1987: 157–8)

---

seeks to justify the Biodiversity Convention: it is seen as overcoming the polarity between North and South; the treaty is lauded for recognizing "interpendency", otherwise unrecognized. But would it not be better to have no more economic development? Does a modern economy destroy nature? The Brundtland report overcomes the polarity by positing that, when development is harmonious, nature necessarily benefits. Measures are necessary, since finance must be distributed properly, but more new practices can again resolve old contradictions: development need not be a war waged by society against nature. The arguments are neat, logical and they always refer to measurable quantities. Biodiversity argumentation suggests professional solutions to global problems. The solutions vary, but the effect is conversational: they share the scene and they tell a similar story. Overcoming the polarity is always dramatic: a contrast is constructed, and then deconstructed to yield unity. But unlike with sustainable development, here the overcoming is pragmatic, rather than comprehensive: new practices are made available, to overcome the conflicts. Overall, the arguments imply cultural renewal, a new scientific culture that will relate differently to nature, a new ethic that will develop. But can cultural renewal be implied, rather than asserted? Can the quiet rhetoric of biodiversity compel attention to force through change?

Holism: Earth

On "Earth", a holistic form of argumentation is developed that relates to the tendency to present a common interest, to overcome conflicts and polarities. This is shown in Box 6.9. Thus, we have the development of a globalizing environmental discourse and a spread of common interest justifications in proposing many other common paths between environmental oppositions.

Each time, the term "Earth" or "planet" suggests an affinity with synthesis, bringing together, bridging, or harmonizing. But this does not preclude differences of opinion. Gore is reconciling human control with respect for the Earth; whereas Eckersley reconciles humanity and Earth, by requiring a new political and social order, a radical reconception. But Lovelock is the most provocative: he uses the argumentative form of "overcoming the polarity" to discuss the relation between human and Gaian interests. The proposal is somewhat surprising: greenhouse pollution serving the Earth!

Concluding on solutions

Throughout this text we have sought to show that environmental arguments are not readily resolved, and that this need not be thought of as a failure of environmental argumentation. Posing different positions is both necessary in view of the variety of actors involved and reflective of a creative interaction between world-views. Conversation, cacophony and debate all show that environmental argument

---

BOX 6.9   Figure: overcoming the polarity

Theme: holism   Topic: Earth variations

Polarity: worship and exploitation; synthesis: stewardship of Earth   "The requirement of stewardship and its grant of dominion are not in conflict;$_{NCD}$ in recognizing
the sacredness of creation, believers are called upon to remember that even
as they 'till' the Earth, they must also 'keep' it." (Gore 1993: 243)

Polarity: planet and people; synthesis: new theories   "To these theorists$_{NCD}$ [eco-
centric emancipatory theorists], a radical reconception$_{NCD}$ of our place in the rest
of nature is not only essential$_E$ for solving our planetary problems; it would also
offer a surer path$_N$ for human self-development." (Eckersley 1992: 117)

Polarity: human interests, Gaian interests; synthesis: common perception
". . . Although another Ice Age might be a disaster for us, it would be a relatively
minor affair, for Gaia. However, if we accept our role$_E$ as an integral part of Gaia,
our discomfort is hers and the threat of glaciation is shared as a common danger.
One possible course of action$_{NPD}$ within our industrial capacity would be the manufacture and release, to the atmosphere of a large quantity of chlorofluorocar-
bons ... They   would   serve, like carbon dioxide, as greenhouse gases
preventing the escape of heat from the Earth to space. Their presence might
entirely reverse the onset of a glaciation, . . ." (Lovelock 1979: 148)

---

is alive and kicking. But the push towards seeking a resolution of the
arguments persists, and some of the most dramatic forms of argumentation relate to the proffering of such solutions. The figures of resolution should not be judged by their ability or inability to close off the
argument. Rather, they should be seen in terms of regenerating the
arguable world of environment and of linking environmental concerns to the broader culture. For modernity is ever in search of the
rational solution, and environmental argumentation on resolution
sheds some light on this search in the modernist tradition.

CHAPTER SEVEN

Rhetoric and culture

The achievement of rhetoric

We have shown how a rhetorical approach can be used to explore the many varieties of environmental arguments and the ways in which they interrelate. Through an emphasis on the active process of arguing, we have furthered the understanding of the dynamics of cultural exchange and the place that discussion of environmental issues holds in the broader culture. Therefore, through providing a rhetorical account of the environet, we have also provided a cultural theory, an account of the contemporary culture through current concerns with the environment. Speaking from *within* the social sciences, Mike Featherstone (1992: vii) claims that such a concern with culture is a radical enterprise:

> The last decade has seen a marked increase in interest in culture in the social sciences. For many social scientists, culture has been seen as something on the periphery of the field as, for example, we find in conceptualizations that wish to restrict it to the study of the arts . . . Culture was too often regarded as readily circumscribed, something derivative which was there to be explained. It was rarely conceived of as opening up a set of problems which, once tackled, could question and overturn such hierarchically constituted oppositions and separations. A set of problems which, when constituted in its most radical form, could challenge the viability of our existing modes of conceptualization.

In our approach, which explicitly crosses the border between the social sciences and the arts, the radical claims of the rhetorical study of culture are even more relevant.

There are of course many ways to analyze culture. Rhetoric has a specific vocabulary for approaching culture through language, and language through argument. The terms are not "technical"; rather, they are ordinary terms made more supple and applicable. Rhetoric dwells within culture and uses its resources, it uses natural inwardness (Gadamer 1979). The key terms that we have used can be reviewed retrospectively.

Argument

Language is seen in terms of argument, making culture a field of arguments. Words make arguments, arguments make dialogues and dialogues resonate to the tune of a culture. The term "argument" has associations of breakdown; by emphasizing dialogue and creativity, rhetoric redresses the balance. Argument becomes the sign of cultures thinking (Billig 1987).

Rhetoric interprets language in terms of voices in dialogue. Voices imply other voices; a proposition implies alternatives. These are understood through the culture that provides the alternatives that "everyone" realizes are implied. Dialogue is sometimes direct, but often oblique, with voices overlapping without being aware of each other: they address a common topic, they contest a word. Dialogues do not "resolve", but they are dynamic. They produce meanings that absorb and reflect ways of life. The boxes in our text dramatize these dialogues of argumentation. "Population" is not an item in a dictionary; it is a terrain of voices, Malthusian prophecy to anti-Malthusian hope, technological remedy to ethical warning. Knowledge is garnered, indeed it is stimulated by the arguments. But knowledge is always cast in words, and the words are always inside an argument, even authoritative words or factual statements. Environmental arguments are contemporary culture thinking about itself and the future, about humanity and nature, about distribution of access and impact.

Discourses

Discourses connect arguments. Therefore, discourse is always about difference between people, between positions. Culture comes together around these points of difference and therefore discourses are argumentative resources connecting people to a culture, unifying "us"

(Billig 1995). Argumentative discourses are available for use in different situations and occasions, linking different worlds. But in a contemporary society, an occasion or situation is not like the Greek polis, the agora or the Roman forum. They are abstract, repeated, and fluidly reproduced in different media (Haraway 1991, Giddens 1991). Imagine the "occasion" of the Exxon Valdez, of the Earth Summit, of Chernobyl: where does it stop? There are no clear boundaries to argumentative occasions in contemporary cultural life. Discourses are the links across such occasions.

Environmental arguers want impact and authority, and therefore they use discourses to meet perceived criteria for argument. But these criteria are always arguable themselves; they can never be settled. New concept discourses adjust to criteria of being comprehensive or Darwinian, experimental or statistically systematic, ethical or professional. New information discourses adjust to criteria of correct sources, diverse references, authoritative data, sophisticated techniques or personal witness. New practice discourses react to demands for relevance, strategy, wide participation or professional expertise, scientific expertise or community involvement, latest technology or traditional resources. Our discourses do make us part of culture, but they do not give us a firm foundation. Culture is not a platform, a support; it is as mobile as the voices that speak it. And those voices provide tremendous repetitive energy for the cultural dynamic.

Differences and realignments

Culture is about interests and desires, hopes and demands, convictions and beliefs. Interests do conflict; desires are incompatible with each other; hopes rule each other out; beliefs challenge each other. Culture therefore inevitably includes conflict, but the conflict is not stable. There are not always two "sides" to environmental argument, such as development versus sustainability, exploitation versus conservation. The "sides" fluctuate and there are moving armies of antitheses. Differences reorientate around each topic. On "resources", there is scarcity against technological optimism, against cultural adaptation, against new economic thought. On "global warming", there are differences over measurement, over probability, over science; and then there are differences over institutional response, over responsibility, guilt and reparation. The opposing views encounter each other within the culture, and sometimes there is focused "debate", with criticism facing counter-charge or defence, as on a pollution incident or an energy technology. Often there is a divergence, the presentation of alternative languages and world-views: Gaia or Darwinian biology, steady state or growth. Sometimes there are islands of conversation, interaction within a consensus, with a common ground of language and world-view: refining the application of sustainable development to different professions. One metaphor for all these differences is *dispersed polarization*. There will always be contrary views, but there is never a fixed agenda to arrange the contrasts neatly.

Opposition invokes conflicts of interest but interests do not determine the presentation of difference through language. Neither do ideas float free. Ideas exist in words and words are ideas. These are used by individuals, each with their own context, their own experience and their own personality. The words also rebound across the culture, taking new meanings as they pass among voices, and they are used in situations of difference. Our aim is not to reduce argument to the static expression of a set of differences associated with a few figures and forms of argumentation, nor to conclude that there are endless and contextless "bubbles of thought" or expressions of environmental positions. Rather, we examine how many nuances are possible from a few figures and forms, given the nexus of interests around the environment, the common exploration of culture through language, and the centrality of the individual in making that exploration. We are actively engaged in the "theatre" of difference (Geertz 1973).

Facts and values

Values are central to cultural dynamics. Culture is about values, and values depend on language (Connor 1992). Values are therefore socially communicative, not a matter of private preference. Furthermore, values are intrinsically arguable, values exist to be contested (Gallie 1960). As language transmits arguments, at the same time it transmits values and therefore at stake is the cultural contestation of values.

In setting values against each other through language, whole worlds are contested. For values are lived out through experience (Williams 1981). This is true for environmental arguments. Key questions about values arise in the discussions: when is the world well peopled; when is a society at home in nature; when is the future

The achievement of rhetoric

secure, and for whom? However, "values" sounds "soft", compared to "facts", to "knowledge" and particularly to "science". The counter-view is, therefore, common: there are facts, and *then* there are values: the fact/value dichotomy (Putnam 1990). Rhetoric has challenged this dichotomy (McCloskey 1994) and our analysis shows how the language of environment compounds facts with values. Population arguments are about data; they also presume values, and values are linked to ways of life. The data inhabit the values; the arguments rework the facts to alter the values, and challenge the values to reinterpret the facts. To cite a classical authority, Mill (1974): the facts do not tell their own story. Environmental rhetoric is about the ways arguments and facts collude, and re-create each other.

Theory as metaphor

So, rhetoric analyzes culture, by interpreting language, and the language of environment is a language of contemporary culture. Language represents culture and it embodies culture. Rhetoric is about understanding how a culture represents itself, and in so doing represents the world. This "fits" with the view of contemporary culture that social theory provides.

The new rhetoric of environment fits with key contemporary theories by way of certain metaphors: never-ending, probing, open-ended, interconnected, uncertain. Metaphors are necessary to theories, as to other discourses (McCloskey 1990); and it is through the *metaphors of process* that rhetoric links to social theory. Rhetoric provides a cultural dimension to social theories, and particularly the social process theories of Anthony Giddens and Charles E. Lindblom. Their theories tend towards social and political structures; our rhetoric presents the implied culture, the cultural possibilities.

"Culture" is of course itself a metaphor, a metaphor for social thinking and the communication of meanings. Linking rhetoric to social theory yields two metaphorical themes: the tendency towards the "answer culture"; and the polemic for a culture of argument. The next two sections explore these themes, concluding with some implications for intellectual enterprise and the story of modernism.

The answer culture

The "answer culture" is a metaphor for certain expectations: that knowledge will save us. The answer culture is inescapable today; it is part of our inheritance. In many ways it is admirable, necessary, and encouraging. It can be seen in many of the more positive statements within the language of environment. Yet it is also at its limits in the language of environment. The culture of argument is a possible future developing out of the answer culture. The culture of argument, if it appears, will be a descendant of the answer culture, not its antidote or antithesis.

The principles of the answer culture are fourfold. First is "the information principle": that it is possible to convert the world into information. Furthermore, when we have better information, we will make better decisions, and our problem is to prevent obstructions developing between the information and the decisions. The second principle, "the therapeutic paradigm" asserts that knowledge makes us better and that this is the purpose of knowledge. Thirdly, "the managerial paradigm" claims that the world is manageable and can be improved by good management. And fourthly, there is "the professional rule" that modern competence is professional, with skills and expertise in the monopoly of the professions.

These principles are inescapable in modern times. We all have a stake in the success of knowledge. But they are not immune from criticism and from the quest to develop beyond them. The question is: how do we interpret its limitations, limitations of which many are increasingly aware (Rosenhead 1989)? And where do we go to develop the answer culture? The environmental arguments show the continuing life of the answer culture, but they also show its need to find new principles, and in the end to take a new shape.

In environmental arguments we find many claims for the answer culture. We see claims from the expertise of environmental science, environmental economics and environmental planning. There are calls for the extension of rational techniques of environmental assessment, environmental auditing and environmental management systems to collate information and inform decision-making. Through this runs the revival or continued reinvention of rational planning method, the means–end model or procedural planning theory (Rydin 1993). Variously formulated, these methods recommend staged sequences of decision-making: survey, analyze, plan; or set objectives, find mechanisms, institutions and tools, implement, and monitor the resulting outcomes. Collect the baseline information, follow the method, train or buy the appropriate expertise, and the answer – the environmental solution – will follow.

Yet this approach has always been open to criticism. The main criticisms have identified: the limits to collecting information, the ways in which objectives become reformulated during the planning process, the need to set arbitrary spatial and temporal boundaries on decision-making, the sociologically bounded nature of actor's decision-making, and the limited powers for implementing decisions in practice. And yet again the answer culture in environmental planning survives. This is because, in discussing environmental planning, in calling on and even challenging the answer culture, the key theme is enlightenment.

The enlightenment is a story about social transition: from tradition, through enlightenment to the knowledge society (Barnett 1995). And now we are concerned with the retelling of the enlightenment story in relation to the environment, just as we approach what may be the end of modernity. Lindblom (1990) starts with the "the Enlightenment's faith in reason" (ibid.: 214). The enlightenment hoped for "a scientifically guided society" (ibid.), where knowledge would set the agenda and solve the problems: the answer culture. Its theorists promoted "universal enlightenment" (ibid.: 220), by which they meant "a radical substitution of reason for power" (ibid.: 221). Pre-modern society had been governed by tradition. Modern society was guided by knowledge. Of course, problems arose, but there must be "faith in reason to solve social problems" (ibid.: 226).

Anthony Giddens also reconsiders enlightenment, "the close relation between the Enlightenment and advocacy of the claims of reason" (1990: 40). In the enlightenment story, natural science has usually been taken as the pre-eminent endeavour distinguishing the modern outlook from what went before. Modernization was the social expression of the scientific spirit and the cultural agenda was set by the success of science. But by now, the answer culture needs saving from itself. It is collapsing under the weight of its own expectations. The answer culture is endangering its own inheritance, unless we find a new adjustment, a new balance between our continued hopes and the repeated disappointments.

Rhetoric illuminates the reasons why the answer culture needs radical rethinking, as demonstrated through the language of environment.

The plurality of contemporary knowledge

In contemporary society, knowledge is plural. There are many kinds of truth, sciences from a self-transforming physics to new psychologies, new technologies of knowledge, and also everyday expertise of all sorts, professional and domestic. Take any issue: how to bring up children, or what to eat for dinner. Imagine how many forms of knowledge vie for assent and implementation. A tanker spills oil into a pristine sea. Which voice counts most: the oil expert, the biologist, the economist, the anthropologist, the witness, the local inhabitant, the campaigner, the reporter? "Which knowledge counts?"; that is the unanswerable question. Knowledge is endlessly unresolved. Thus, no moral can be drawn from scientific advancement: "biodiversity" is a scientific concept, but it is also about the negative impacts of science, and it implies a world that will be forever unknowable. Biodiversity is a triumph of measurement and, after the measuring, a validation of impersonal method, but also a return to local understanding.

In the environet, modern culture begins to outreach itself. We necessarily leave the politics of the answer culture, where knowledge was authoritative:

> Politics cannot be reduced to expertise; but also expertise can never sustain the claims to legitimacy possible in more traditional systems of authority. (Giddens 1994: 95)

Perhaps the most profound indicator is the *ethos* struggle. Environmental ethos is about knowledge and trust, but there are now many ways to claim trust: from reassuring realism to essential vision; from lofty far-sightedness to angry objectivity; from alarmed knowledge to knowing mockery. The ethos of knowledge is diverse: it varies within topics and across the environet. The ethos of essential vision fits sustainable development, the ethos of far-sightedness fits global warming. In some contexts, such as "population" or "rainforest loss", objective ethos is heightened by emotion, but then the emotion invoked can vary. Consequently, ethos is diverse, even within a text, if it attempts to cover different topics. This is one reason why the great environmental text is not going to be a modern, consistent document but, if anything, a postmodern play of different voices.

The complexity of contemporary "problems"

For Lindblom, the enlightenment was a story about social problems and how to solve them, a story that is no longer plausible. It cannot be told straight any more, there are too many ironies and twists, too much suspense. Lindblom concedes "a degree of retreat from the Enlightenment" towards a society that "pursues inquiry and the resourceful utilisation of its results more than it pursues firm knowledge" (1990: 301). Lindblom goes beyond conceding that answers are unattainable; he argues that we should not be seeking "the answer", we should instead be pursuing an inquiry. Too many A's have returned as not-A's, to re-apply Kenneth Burke's view of irony: convenience as poisoning, wealth as alienation, benevolence as tyranny.

Giddens also dramatizes the irony of late modernity:

> Events have not borne out these ideas. The world we live in today is not one subject to tight human mastery – the stuff of the ambitions of the left and, one could say, the nightmares of the right. (1994: 3)

Knowledge has not delivered the answers to our problems. On the contrary, our problems seem to have developed resistance to our knowledge:

> The uncertainties (and opportunities) it creates are largely new. They cannot be dealt with by age-old remedies; but neither do they respond to the Enlightenment prescription of more knowledge, more control. (ibid.: 4)

Tradition will not work, but nor will discovery save the world. Our problems are *resistant* to knowledge and its answers. We must keep trying, and so, in such a society, "new" discourses multiply: new concepts, new information, new practices (and new rhetoric). But the "new" is plural, endless, and indefinite. One irony in particular is inescapable, an irony Giddens terms "manufactured uncertainty":

Manufactured uncertainty refers to risks created by the very developments the Enlightenment inspired – our conscious intrusion into our own history and our interventions into nature. (ibid.: 78)

Our knowledge has actually created problems, riddles, and puzzles. Our control has implied dilemmas.

The fund of questions

No matter how good the data, people will also ask: should I believe this kind of argument, who is trying to convince me, and with whom (as well as what) would I be agreeing? After two centuries of enlightenment, modern culture remains a hubbub of questions. For every answer, there are three questions:

Questions abound about what to believe and what kind of evidence, if any at all, to require as a condition of belief, as well as practical questions about how to go about structuring questions to be attacked. (Lindblom 1990: 34)

Evidence is amassed; but that provokes the question, what evidence should we count? The more the evidence, the more acute the question; whose evidence, which rules apply? The answers come into focus, but then the problems are redefined. Rhetorical topics are a good metaphor for this process: they are not soluble either, only discussible. And the discussion remakes the topic: as we discuss it, the topic changes shape. A key example is sustainability and sustainable development: "Who am I to believe?". No amount of refinement will answer the question: no amount of definition will prevent people questioning. Do we actually want to believe someone so definitive, they will ask? Another example is Gaia: "What kind of evidence is required to convince us, will computer simulation do, are statistics relevant, are visions acceptable?"

We are living through the last stages of "modernity's repression of existential questions" (Giddens 1994: 217). Existential questions concern "how shall we live?" (ibid.: 212), and for Giddens ecology is an existential agenda:

Ecological problems disclose just how far modern civilization has come to rely on the expansion of control, and on economic progress as a means of repressing basic existential dilemmas of life.

Ecology's questions are basic; at issue is our way of life itself. Knowledge helps pose the questions, but it cannot give the answers. We will have to work out how to live some other way.

The arguability of science

Science repeatedly enters our environmental arguments. But instead of settling decisions, science itself becomes questionable. In the environet, we have passed the culture of "simple modernization, where science and technology were supposed to generate unquestionable truths." (Giddens 1994: 95). Instead as we discuss global warming and resource depletion, population and energy: "The findings of science are interrogated, criticized, made use of in common with other reflexively available sources of knowledge." (ibid.: 216).

Recognizing both theoretically and practically that modern science is socially constructed (Taylor & Buttel 1992, Murdoch & Clark 1994, Wynne 1994, Yearley 1993), means that science is questioned rather than accepted: it is this questioning that takes us definitively beyond the answer culture. Indeed, science keeps changing, and there are different sciences in competition with each other: to accept one claim is to question others. Time and again, we encounter "regular shifts in knowledge claims" (Giddens 1991: 123) – experts say global warming is accelerating, but others say it is not clear; experts have said sunshine is healthy, then they say it is dangerous. Cumulatively, science itself is contested and contestable.

The answer culture is on the defensive and retreating. But within the environmental topics the mood is far from anti-science. On the contrary, the claims for science are diffused; science is everywhere, it takes multiple forms. The answer culture is undermined by the multiplication of science, not by its dissolution. Science has always meant different things, and environmental argumentation derives partly from the arguability of science. We are left in the rhetorical terrain, where "not only officials but ordinary people make use of information and discussion in formulating problems and the positions they take on problems" (Lindblom 1990: 3). Science helps us make our arguable worlds, but it cannot release us from argument.

There are still many attempts to define environmentalism with

BOX 7.1   Science and the environet

Mainstream science means mainstream environmentalism   ". . . dry greenism is only a comforting start of a process of more radical ideological and structural change. In the middle is shallow greenism, the fashionable centre ground of most Liberals, the bulk of environmental scientists and far sighted industrialists and academic economists." (O'Riordan 1993: 307–8)

New science means radical environmentalism   "Deep greenism takes a more radical view of the world. It draws its perspective from a series of loosely connected ideas that depart from mainstream science, economics, politics and philosophy . . . Humans are marginal to the real wellbeing of the Earth . . . communal self reliance . . . social fundamentalism . . ." (O'Riordan 1993: 317–18)

Pro-science sustainability versus anti-science environmentalists   ". . . We may note a difference in emphasis between advocates of sustainability and some modern day environmentalists. The former must believe in technological advance, whilst not being so simple minded as to regard all technological change as 'benign' (we know it isn't – think of CFCs). The latter also tend to be anti-technology." (Pearce 1993: 5)

Environmentalism means new interdisciplinary science   "The more I read of her book, the more I admired Henderson's trenchant analysis of the shortcomings of conventional economics, her deep ecological awareness, and her broad, global perspective . . . As I carefully mapped out the multiple interconnections between economics, ecology, values, technology and politics, new dimensions of understanding opened themselves . . ." (Capra 1988: 248–9)

Political environmentalism is anti-technological, anti-scientific   "But the Green movement is itself a potent force preventing environmental reform. It has anachronistic views of industry, which it regards as harmful and polluting." (Lovelock 1992: 111)

Environmentalism is anti-scientific fundamentalism   "Check out Moslem fundamentalists or the right-to-lifers. Like abortion opponents and Iranian imams, the environmentalists have the right to tell the rest of us what to do because they are morally correct and we are not." (O'Rourke 1992: 22) regard to science. But the definitions (see Box 7.1) are themselves cacophonous, expressing the ambiguities of science as well as of environmentalism. The definitions exemplify a post-enlightenment condition. Science is an arguable setting for environmental arguments, always a reference point, whether positive, ambiguous or negative, always at stake. The idea of science is made mobile and indeterminate, essentially by a series of associative arguments. One association links "the bulk of environmental scientists" to "shallow greenism", middle ground environmentalism. Another links new science, beyond the "mainstream" with "deep greenism". In another antithesis, anti-technology positions go with extreme environmentalists, as against sustainability advocates. For still another associative argument, "economics, ecology, values" move together into a new scientific vision. But Lovelock reassociates the whole green movement with anti-technology, the hatred of scientific progress, whereas O'Rourke associates environmentalists with religious fundamentalists. And so the spectrum is complete, from environmentalism as part of mainstream science, through environmentalism and new science, to environmentalists and anti-scientific beliefs. In the process, environmentalism and science are equally contested and ambiguous.

So, on the environet, science enters the fray about resource depletion and environmental degradation, and the result is a new argument, about the role of the latest knowledge in public culture (Box 7.1). Science supplies the concept of biodiversity and its powerful new information discourse; it also supplies the new concept discourses of Gaia. The one glances back to Darwinian theory; the other is opposed by major modern Darwinians. Contemporary discussion transfigures pronouncements into propositions, monologues into dialogues. The harder we try to settle an issue, the more it rebounds.

The answer culture is a driving force, a driving force inside the language of environment. The rhythms of environmental language are the rhythms of the answer culture; the perpetual drive for new concepts, new information, new practices; the demand for resolution, mirrored in so many inventive figures of resolution; above all, the ethos of knowledge diversified in so many new ways. But the whole effect is a cultural paradox: the answer culture outgrows its own motives. There can be no answer when there are so many possibilities. There can only be alternatives, better and worse. And one alternative might be a different culture.

A culture of argument

The goal of post-enlightenment democracy

There is a positive reading of the environet that is possible, pointing towards a new democratic culture, a culture of argument. The environmental arguments have succeeded where they least wanted to: they have helped create the grounds of a new culture of argument, rather than a new consensus.

> A competition of ideas offers some escape from impairments . . .
> You or I could enjoy at least a limited opportunity to form a reasonable judgement not wholly crippled by impairment, say, about the prospects for global warming or improved day care facilities for infants because the many people and institutions that attempt to control us – say to win us over to this or that organization, party, or candidate – dispute with each other and give us the benefits, unintended, of their challenges to each other. (Lindblom 1990: 81)

The benefits are unintended: each party tries to win the argument, wanting the discussion to finish. But the result is ironic: the discussion takes off on new tracks. Rhetoric also tells the story of such an irony: the more texts pronounce environmental topics closed, the more the topics are alive. The classic examples are the figures of resolution. They seek to end the arguing and yet they are some of the most lively arguments. One synthesis counters another; dialectic of catastrophe is countered by feasible possibility, and one final feasibility by an alternative. Lindblom refers to "epiphenomenal problem solving" (ibid.: 244), "by which decisions or acts systematically produce betterment or solutions intended by no one". Rhetoric uncovers how epiphenomenal problem-solving happens, and it explains what a culture of epiphenomenal problem-solving might be like. The arguments are simply too many, there cannot be a solution directly arising from them. All arguments have so many rivals that they could not possibly say whether they have won or not. But the process of arguing is itself creative, it delivers new ideas, new facts and new remedies, unpredictably and spontaneously.

Contemporary culture demands new ways of thinking, and particularly new ways of thinking about "unsolved" problems. We need to outgrow the old models, the models where truth comes to the rescue, where the viewpoints are gradually eliminated until only the right answer remains. Only repression will remove the questions expressed through the environmental topics: otherwise, there will always be more differences, new alignments. Rhetoric helps to rethink "problems", to see them as arguable worlds we live in, rather than puzzles we design and dissolve.

If the goal is argument over possible worlds, what is the role of consensus rather than difference? What help would consensus be? Lindblom has a moral for social theory "we must reconsider . . . How a society or polity holds together." Agreement is not the glue:

> We turn from celebrating agreement, consensus, and harmony as sources of social order, social peace, and democracy, to facing up to them no less as cripplers of minds and obstructions to using them for human betterment. (ibid.: 129)

Agreement is suspect: there are *always* other ways to put the case, other perspectives. Lindblom sees consensus as the other side of "impairment", the injured inquiry, deprived of energy, encouragement, and information. All inquiries are impaired (ibid.: 78–80): the information is never neutral or complete, the institutions are never impartial. But we cannot afford to wait for the perfect context, when knowledge will finally be free and fair. Instead, we must pursue our differences and keep probing. So, it is not a bad thing that the envi-ronet is endless, and that realignment in difference is the response to every innovation, realignment rather than consensus.

But people expect arguments to be resolved. If the argument is perpetual, it is disturbing, it seems to be "pointless". And there *are* dangers: that environmental discussion might tire people out, that the interest might be fading (Bramwell 1994). Perhaps environmental culture could come to the rescue, even as environmental politics stands accused of failing to deliver.

> A frequent social science error holds that when the volitions of various people or groups of people conflict with each other a reconciliation must be imposed. In the face of interpersonal conflict, what else is there to do? But, of course, there exists an "else" . . . More probing, more discussion. At any point at which conflict

exists, more inquiry and more mutual persuasion continue to be a possibility. (Lindblom 1990: 47)

More discussion justifies itself: we do not *need* everyone to agree at the end. There is material enough for an endless "else", an assurance that there is something else to do. Discussion proceeds; sometimes a problem will be met firmly; at others, it will be replaced by a new formulation. Truths *will* be generated: the arguments will not be satisfied by mere semblances. They will need the latest sciences, the latest data. But there is no reason why people should then use the data to agree with each other. A link is therefore broken, the link between knowledge and agreement, science and consensus; this is the link of enlightenment itself (Habermas 1984, 1987).

Following the cultural turn

So, a new goal of the argument culture is made conceivable now, a goal that promises a post-enlightenment democracy. But there is always the question of *how* we move towards this goal. Clearly, some institutional change is necessary to achieve the goal, but changing organizations is not sufficient in itself. The oil of cultural argumentation is needed to make organizations work, to generate support for change and put change into effect. Identifying how cultures should move in response to a programme of institutional political change is not easy. Cultural change does not readily map onto the politics and institutions of democracy.

Lindblom provides some terms that bridge the move from democracy as politics to democratic culture, following through on the cultural turn in politics. The key term is "reconsideration":

If they do not perish through catastrophe but work their way to a self-guiding society, they will not do so because professional inquiry has discovered the route to it for them. Given their constant reconsideration of both ends and means, there exists no route to be discovered, only routes they must create. (1990: 302)

The key metaphor of enlightenment, discovery, is pushed aside. Instead there is "reconsideration", and "re" is the key syllable. "Reconsider" is what you do when you have made a mistake, or when you aren't sure any more: "please reconsider, before it is too late,

A *culture of argument*

before you go any further". Reconsideration is the antithesis of discovery, indeed it has long been the rival metaphor. *Reconsideration is the metaphor for the environet*. Our rhetoric supports a shift of metaphor, from discovery to reconsideration. It belongs with the reconsidering stories, not the stories of great leaps forwards. But will they reconsider before it is too late – "If they do not perish through catastrophe . . ."; the story does not and cannot tell. This is a central point for environmentalists to grasp.

Reconsideration is also allied to reflexivity, a key term for Giddens:

> Decisions have to be taken on the basis of a more or less continuous reflection on the conditions of one's action. "Reflexivity" here refers to the use of information about the conditions of activity as a means of regularly reordering and redefining what that activity is. (1994: 86)

Reflexivity is our condition. And Giddens offers new interpretations of democracy for a reflexive age. First, there is deliberative democracy. The key to deliberative democracy is openness, rather than seeking to overcome all differences. It implies a sophisticated approach to conflict resolution.

> The deliberative conception of democracy . . . Accepts that there are many questions either which have no single correct answer, or where solutions are thoroughly contested. (ibid.: 113)

Decisions must be made but without pretending that differences have been smoothed away. The differences may return and the result is the perpetual democratization of public discussion. "Democracy in this conception is not defined by whether or not everyone participates in it, but by public deliberation over policy issues." (ibid.: 114)

Then Giddens moves on to "the democratization of democracy" (ibid.: 114). Democracy goes beyond formal politics to address how we live; "dialogic democracy" is the democracy of relationships and groups. The principle is dialogue, but not dialogue in a conventional sense, talking that can resolve differences. On the contrary, talking is not about resolving all differences, perhaps not even many differences will be resolved by dialogue. There is a link with rhetoric.

Dialogic democracy is cultural, not political, and implies a radical shift in culture:

> Dialogic democratization is not an extension of liberal democracy or even a complement to it; . . . Dialogic democracy is not primarily about either the proliferation of *rights* or the representation of *interests*. Rather, it concerns the furthering of *cultural cosmopolitanism* . . . (ibid.: 112)

The dialogues are not about rights, claimed rights and opposed rights. They are not about demands and rejections. Nor can the dialogues be explained as expressing interests, fixed and clear cut group interests. Of course power is at stake. But the dialogue cannot be interpreted through the medium of traditional political debate. It is not focused by clearly opposing ideologies and their interests; it does not tend towards a great conflict. Instead, the differences multiply, diversify, and the conflicts are local, plural, and ambiguous. The result could be "cultural cosmopolitanism". "Cosmopolitan" suggests an atmosphere, an orientation: accepting different worlds, intrigued, curious, ironical. Cultural cosmopolitanism would be at home with differences, the play of difference, unresolved and unresolvable. There would be no need to achieve consensus or conformity. "Dialogic democracy . . . doesn't imply that all divisions and conflicts can be overcome through dialogue – far from it." (ibid.: 115)

By applying "reconsideration" and "reflexivity", Lindblom and Giddens look towards the democratization of culture. They ask, to take Lindblom's inflection: will society become self-directing? Are we capable of outgrowing the image of "the science society", where knowledge solves all problems, where someone else's knowledge solves *our* problems? Will we acknowledge the real questions, the "basic existential dilemmas", or will we continue to suffer the consequences of their repression, the refusal of space to think and feel them?

The environet is a sign of the strain of transition but also a hint of hope. From the point of view of "the scientifically guided society", there are too many arguments, and we need to reduce them. But from the point of view of an *inquiring* society, we need all the arguments we can get, we need the dilemmas to be exposed. The environet could be an early prototype of the agendas of a culture of argument, agendas framed such that the old solutions are inconceivable. Environmental rhetoric is therefore a study in cultural reflexivity, to take Giddens' term, or in open-ended reconsideration, to adapt Lindblom: both point towards a culture of argument.

The environet links politics and experience, information and feeling: it could portend an inquiring culture, emerging from the information age, indeed emerging at the vortex of the information age. Dialogic democracy would favour the spread of the environet. But there is the possibility that the environet might come to signify the triumph of information, the overwhelming pressure of expert claims. The existential dilemmas may emerge, come to expression; or they may be repressed afresh, by the renewed claims of all the solutions. The story has yet to unfold in full.

Contemporary politics and environmentalism

So there are hopes and strains within the environet portending a new form of democracy. But how does this relate to contemporary political activity? Commentators on environmentalism have been keen to develop spectrums of environmental thought (O'Riordan, 1992, Dobson 1995) from light to dark green, from ecocentric to technocentric, as if from left to right, although acknowledging that one spectrum does not simply map onto the other. The problem is that the political map is unstable: that is why we have been able to show such complex differences and realignments over environmental issues.

Giddens asks, "Could one see in the extraordinary upsurge of green ideas over the past few decades the springs of a renewed political radicalism?" (Giddens 1994: 198). His answer is "yes and no", as it were. If environmental politics is radical, is it left wing? Not quite:

> Yet although green movements often tend to situate themselves on the left, there is no obvious affinity between radical ecology and leftist thinking. (ibid.: 200)

If environmentalism involves conservation, is it right wing? Certainly, one can note:

> . . . the affinities between green philosophies and conservatism . . . However, ecological ideas do not have a privileged connec-

tion with conservatism any more than they do with the left or with liberalism. It would be more accurate to see green philosophies as reflecting the shifts in political orientation . . . (ibid.: 201–202)

The associations spread outwards "beyond left and right" in Giddens' metaphor. "Left" and "right" have been the metaphors of modern politics; whether we are going to need new metaphors is arguable. Box 7.2 shows how environmental politics is interpreted using the modern political grid. But the problem is to redefine politics, not just to define environmentalism. Environmental arguments go on within politics and at the same time, reflexively, they redefine "the political".

The intellectual challenge
The rhetorical analysis of the environet offers new possibilities for intellectual as well as political practice. We frame these in terms of three principles: epistemological patience, methodological spontaneity and lucky thinking.

Epistemological patience This environmental rhetoric dramatizes the explosion of the answer culture, showing how information creates uncertainty, how new insight generates new ambiguities. No-one wins the argument, because there is not one argument but many: "it seems better to emphasize the activity of inquiry than take stock of findings." (Lindblom 1990: 35). A new state of mind is therefore needed, a kind of dynamic patience, patience in the face of plural and overlapping knowledge, epistemological patience. The topics are not immature agreements, awaiting their full development as knowledge; they are growing inquiries, which will never outgrow their "open, never-ending and inconclusive character" (ibid.: 34). We must stop waiting for our discussion to grow up, and start to live with "the open-ended and exploratory quality of thought" (ibid.: 35). Every move creates new arguments, new differences always accompany new agreements.

The outcome can be seen in different ways. Yes, it does illustrate that knowledge does not turn up at last with the answer. No-one agrees when the data is finally announced, the latest report on climate change or on population trends. Global agreement inaugurates new

BOX 7.2 Politics and the environet

"Get real": environmental politics goes official   "What's got to come now is environmental politics; we've got to green the politics of this country so environmentalists can get into office." (David Brouwer in McKibben 1992: 242)

"Join with us": environmental politics means new community   "And if any radical movement for social change and an ecological balance between second and first nature can be achieved, it must be based on a participatory democracy, rooted in a politics of gradual confederalism – the step-by-step formation of civic networks." (Bookchin 1991: liv)

"Right on": environmental politics and new rights   "The new environmentalism . . . will prevail also because it is locked into other social movements, notably those pertaining to rights to health, rights to know, consumer protection, feminism, pacifism and animal welfare." (O'Riordan 1993: 297)

"Dinosaurs": environmental politics and the "Old Left"   "By virtue of the radical democratic and participatory nature of environmental protest in the 1960s, many political commentators tended to regard it as an adjunct of the New Left. Yet even this association was soon to come under challenge as a rearguard action developed against environmentalism by labor, socialist, and liberal welfare activists and theorists." (Eckersley 1992: 10)

"Grassroots": environmental politics is grass-roots politics   "Some mainstream environmentalists charge that minority grass-roots organizations aren't sufficiently concerned with the environment, that they seek only a fairer distribution of environmental dangers – 'environmental equity'. . . . Such allegations, however, miss the radical potential of grass-roots environmental activism." (Rosen 1994: 229–30)

"We the people": environmental politics is new global politics   "On one side are scientism, developmentalism and statism, backed by the big battalions of the establishment: modern technology, and the institutions of world capitalism and state power. On the other side are the people, principally the 30 per cent of humanity that is disposable as far as the modern project is concerned, but aided and abetted by many from the 70 per cent who regard this project as ethically, socially and environmentally intolerable." (Ekins 1992: 209)

disputes, as on biodiversity. But on the other hand, more knowledge arrives, ideas do turn up, and there is no end to the good arguments.

Methodological spontaneity   Answers are plentiful. They turn out to be easy – as long as one realizes that there are always alternatives. It is the questions that are hard: the trick is to discern the necessary question. Topic analysis is about the cultural power of questions. Lindblom's stories are about "probing", "the varied, open, never-ending and inconclusive character of . . . investigations of a social world in motion" (1990: 34). Probing is normal. Everyone probes, not only specialists. People keep coming up with new questions. If "methodology" is useful, it must refer to the rich reinvention of questions, and there is no one way to think of good questions, no high methodology:

> One also underestimates the richness of everyone's inquiries by overlooking the pervasiveness of methodological inquiry of many kinds.

Academic methodology actually lags behind everyday methodology: people are good at thinking up questions – that is the state of contemporary culture, as it passes beyond the answer culture. The questions are often rhetorical; for instance, the question of ethos naturally arises, when faced by knowledge claims: "A probing mind does not merely ask if acid rain despoils forests, but also asks, "Who am I to believe about acid rain?" Rhetoric connects academic inquiry to "the richness of everyone's inquiries". It is therefore time to endorse "the varied, open, never-ending, and inconclusive character" (ibid.) of inquiry, to conceive of methodology, if at all, in a new sense: methodological spontaneity.

Lucky thinking   In the answer culture, the right idea is always expected, and it is a matter of organizing ourselves and our institutions to find it. After the answer culture, some ideas strike lucky, they hit the moment:

> The science model society tends to regard society as a formal organization which recognizes problems and then assigns problem solvers to them. By contrast, self-directing societies leave a

great deal of room for epiphenomenal problem solving, in which solutions to problems emerge not from deliberation or design but as by-products of people's attention to other concerns or problems. (ibid.: 223)

Ideas need luck. The problems are entangled, enmeshed in each other, susceptible to differing interpretations. We will not solve the problems by sitting down the best experts and letting them find the answer. Which experts? Which version of problems? The environet is an example of a new way to formulate "social problems". Under the old model of the answer culture, these problems would be insoluble. No-one could get a grip on all these ideas, no one could devise a plan that would deal with all these possibilities. Therefore, seen from the point of view of the old model of problem solving, it looks disheartening. Indeed, it does not look like knowledge at all, because it cannot be used to support a solution.

On the other hand, if one adopts Lindblom's social model, then the story is different. The environet is exactly how problems might be represented – not to invite a solution directly, but to make connections apparent, to facilitate lucky thinking.

Social science would also need to adjust to this new problematic. Problems would be revolved, explored, and adjusted; they would not be dissolved. Social scientists would not be "problem solvers"; they would be discussion leaders, "all of which cast professional inquiry in a role supplementary to lay inquiry." (ibid.: 161)

Discussion continues, but no-one waits for the right answer; the better answer will do for now: "In the self-guiding model, the usual test of a good solution is instead that it has been well probed. Competent opinions will differ on whether it has." (ibid.: 218). Rhetoric fits the self-guiding society. The rhetorician is attuned to the arguments. Rhetoric reorganizes topics and reveals other ways to invent cases, other places to find persuasion. It shows where the discussion is lively, where else it is running out. It maximizes the chances of a lucky idea.

Environment and the boundaries of science   So, what are the implications of this new intellectual practice for the disciplinary organization of education, research and thought, for the distinction between natural, social and human sciences? Modern environmentalism inherits a fundamental question about the role of the sciences of society. In their founding work, Ward & Dubos (1972: 290) analyzed natural science via social science:

> There are three clear fields in which we can already perceive the direction in which our planetary policies have to go. They match the three separate, powerful and divisive thrusts – of science, of markets, of nations – which have brought us, with such tremendous force, to our present predicament.

They were critical; they also proposed a new science, a new interdiscipline, and a linkage between natural science and social science:

> And they point in the opposite direction – to a deeper and more widely shared knowledge of our environmental unity, to a new sense of partnership and sharing in our "sovereign" economics and politics, to a wider loyalty which transcends the traditional limited allegiance of tribes and people.

These calls for an alliance between natural and social science are not uncommon in the environmental literature. But the rhetorical approach suggests that perhaps we need a different adjustment – an adjustment between social sciences and lay probing, rather than an adjustment towards a new metascience. In this adjustment lies a hope of a new democratization, a new intellectual agenda and a new culture of argument. The environet shows these possibilities and the pitfalls in moving towards them. Environmental arguments reveal much about the culture within which they exist and the boundaries between cultures.

Environet and modernity

So, to conclude, can the rhetoric of the environet shed any light on the enlightenment, on the trajectory of modernity? Contemporary social theories disagree about the uniqueness of this time (Berman 1982, Bauman 1992, Fukuyama 1992, Baudrillard 1994, Giddens 1994). Social theories are polyphonies, a play of differing voices; the polyphonic themes are novelty, surprise, and shock. Transition stories abound, stories about when the world began to change, how deeply it has changed, what remains of the past. We write from the middle of these stories.

The ending is the issue, of course, as it must be from mid-story. One major storyline foresees the end of history: a social form triumphs, liberal democracy and the market (Fukuyama 1992). Another major storyline is the return to barbarism, the collapse into darkness (Hobsbawm 1994). In between are stories such as the work of Giddens and Lindblom, open-ended, ambiguous to the horizon and beyond. The plot is still evolving; no-one knows the winners and losers. In particular, no-one is sure whether enlightenment wins or loses the struggle with world history. The "end of history" is a victory for liberal enlightenment; the "return of barbarism" is the defeat of enlightenment. Giddens and Lindblom imply that another story is occurring. In *this* story the old enlightenment is over. There is no future society shaped by a grand science, a great theory whether natural scientific or social scientific. Progress is not linear.

The environet is a cross-section through the middle of the story of modernity, the story whose ending is unknown. Most environmental topics contain ironies about progress: global warming from industrial technology; air pollution from modern transport. Yet, in the language of environment, knowledge is promoted. Knowledge will not settle the disputes, but all arguments relate themselves to knowledge, to theories and information. The environmental arguments can be read in different ways. One reading is that progress is no longer persuasive, knowledge no longer authoritative; the other reading is that progress is still worth arguing over, and that knowledge is still worth contesting.

Modernity is the key issue, not postmodernity. The argument is about the modern world, what modernization amounted to, and where it may still take us. Throughout environmental arguments, modernity is a key issue: resources discovered, consumed; plans devised, unhinged; technologies unveiled, exposed. Sustainability may rescue modernity, or it may stake a claim to replace modernization. Steady-state economics is against the economics of growth, and so of orthodox modernization; but steady-state theories are still scientific. The modern story is still unfolding.

The answer culture told stories about knowledge saving the world and modernizing old forms of life. The stories are still compelling, but they can only be possible stories now, not necessary stories. We need to add new narratives about modernity, narratives in which the limits of the answers are not fatal to the hopes of progress. We need to tell other stories about modernity and knowledge, stories in which the answer is partial, or provisional, or contested, and yet in which there is still a possible advance. We have suggested, along with Giddens and Lindblom, that the heroine of the story could be a new democracy, a democracy giving expression to plural enlightenment, a democracy whose principle is open-ended discussion, not single verdicts. It could be a story of enlightened uncertainty, the birth of democratic culture. So, we end with hopeful uncertainty. It is sometimes said that we are entering "the information age" or "knowledge society". But in an important sense, nothing could be further from the truth. Information has never been less able to help us: instead, we must help the information. The moral is that *we* must make knowledge meaningful through our arguments.

Appendix One

Reading environmental texts: the methodology

This project is based on a close reading, within a rhetorical framework, of a large number of environmental texts. Much more material was "read" than is presented here. In order to achieve this, use was made of the new information technology now available in libraries. In particular, we used an on-line database of newspaper material called Profile and a selection of CD-roms.

The Profile database is operated by the *Financial Times*. It is updated daily and it draws on an increasing range of US and UK newspapers and press services. At the time of the research, the US "file" comprised:

- *American Banker*
- Associated Press
- *Business Week*
- Dow Jones News Database
- *Washington Post*

The UK "file" comprised:

- *Today's Financial Times*
- *Independent/Independent on Sunday*
- *Financial Times*
- *Guardian*
- *Today*
- *Daily Telegraph/Sunday Telegraph*
- *The Times/Sunday Times*
- *Observer*
- *Lloyd's List*

- *European*
- *Scotsman/Scotland on Sunday*
- *Northern Echo*
- *Herald*
- *Evening Standard*
- *Daily Mail/Mail on Sunday*
- *Irish Times*

It is possible to interrogate this database interactively. Selections can be made by "file", newspaper, time period, keyword and the presence of the actual word in the text of the press report. We used selection by the appearance of an environment word in the text. To begin with, we scanned across the years 1990, 1991, 1992 and 1993 with over a hundred keywords, sometimes using the word on its own, sometimes requiring it to appear with the word "environment" to ensure relevant coverage. This gave us a *count* of the incidence of environment words and allowed us to select the most central terms. These amounted to about 30 words. We then collected full coverage for these words for a specific period in December 1993; a tight time period had to be selected to constrain the number of references. For these words in this period we printed out the full text of the articles containing the words. This was then read carefully to provide the initial analyses. From this, the final selection of environment words presented in the book was made. The newspaper coverage for the final selection was re-read with a view to identifying quotes for detailed commentary and structuring the final rhetorical account.

This account of the methodology emphasizes that it is a highly interactive process, with selection of the material and refinement of the reading going hand in hand. It also suggests that the computerization of press material can lend itself to the qualitative analysis of the material as well as the more obvious quantitative analysis (Lacey & Longman 1993, Mazur & Lee 1993).

The CD-rom search provided equivalent access to academic and related literature. CD-roms are now extensively used in literature searches by academics and students alike. They allow search by keyword rather than presence of the word in the text, and different CD-roms offer coverage of different selections of journals and books. Generally, these are organized along disciplinary lines. To overcome disciplinary boundaries we used a range of CD-roms:

- *BIDS*: Bath Information and Data Services for sciences, social

sciences and arts
- *Econlit*: bibliography of economic journals
- *Geography*: bibliography of physical and human geography journals
- *Sociofile*: bibliography of social science journals
- *Enviroline*: bibliography of environmental literature
- UKOP: catalogue of publications of UK official organizations
- USGPO: catalogue of US government publications
- SCAD: catalogue of European Union publications

Again an interactive method was used. A first scan was undertaken with the larger number of environment words (135 in this case) to provide a count and assess which were the most salient in this academic context; this fed into the final selection of the key terms. Abstracts were scanned for the interim selection of 30 or so words and a selection printed out, reflecting the *diversity* of discussion around these themes. This printout of abstracts provided a way into the breadth of academic discussion on the environment. From these a selection was made for obtaining copies of the full text of articles and papers. This "personal library" of photocopies was supplemented with a trawl through the on-line catalogue of the British Library of Economic and Political Science (BLPES) "Libertas" and many bookshops and publishers' catalogues to identify new and recent books on environmental topics. We cannot reasonably claim to have read everything thing on the environment, of course. But we do believe our method has allowed us to cut across the huge and growing environmental literature to identify key themes and select not representative or core texts for reading, but interesting and diverse texts.

Again the resulting books and papers were read in two stages: first, to provide an initial rhetorical reading of topics, tropes and figures; secondly, to select quotes and read those again for detailed commentary. Two of the advantages of co-authoring are that while you don't each have to read everything, alternative readings can be made of selected items. And most of the books and all of the selected articles were read by both of us. The extracted material has been re-read several times by us both. Multiple readings are a key element of such a methodology, whether by one person or a group.

And readings have to be turned into writing. We found that it was most productive to begin with a rhetorical account, topic by topic, word by word. This can, however, produce very extended accounts of the individual elements. To conceptualize the interrelation of these elements in the environet required a restructuring of the account. Hence, we reordered our text according to the key terms of rhetorical analysis; in this way, the current chapter structure was devised. This account shows how the reader (and writer) has to be responsive to the original texts **and** to the commentary, the text that is being created. Rhetoric is an endless circle, of reflecting back on the material under scrutiny and on the account the rhetoric developing. Any such qualitative method will be interactive and self-reflexive, as fits the theoretical accounts it derives from and hopefully contributes to.

Appendix Two

"Doing" rhetoric: a nuclear exercise

In this appendix we offer a selection of un-annotated quotes on nuclear energy from two sources. First, this material comes from a selection of more or less polemical, more or less academic texts; secondly, it comes from the newspaper media. We offer no commentary, but invite readers to explore the rhetoric of these arguments themselves. This will involve careful reading and re-reading of the material, paying close attention to the words.

- Which forms of discourse are drawn on: new concept, new information, new practice discourses?
- What are the prevailing alignments and differences of views found in the quotes?
- Consider the prevailing ethos in and across the quotes.
- What about the figures of argument; can you note the metaphors, the associations, the use of irony?
- How does the arguer attempt to close off the argument; what is the figure of resolution?

Nuclear technology was and still is the paradigm of modernist technology, the original "white heat". Therefore, arguments on nuclear energy also provide a commentary of and on modernity. How do they reflect on our contemporary society as it enters the distinctive period of the late 1990s?

BOX A2.1 Nuclear Exercise

"What could not be confidently predicted before Chernobyl were the local effects of such an accident . . . As a result, the whole district had to be managed on something like a 'war footing' and efforts resembling a large military operation were needed to contain the damage." (WCED 1987: 185)

"God could therefore afford to surround us with radioactivity and irradiate us at all times and in all places because the adverse effect upon us is extremely modest. The effect of a nuclear power programme upon us would be even less . . . We know that, at Chernobyl, the population living nearby received on average a dose of 100 milliSieverts (mSv). Now the average lifetime dose in the Ukraine is 200 mSv . . . It is for these fundamental reasons that we can remain confident that the latest scare story from Chernobyl or from Sellafield is just that, a scare story, not something which will survive careful scientific examination." (Marshall 1993: 155-7)

"The purpose of this personal digression is simply to establish my credentials for objectivity in the matter of nuclear power. I have seen the worst that it can do. I nevertheless share Lord Marshall's conviction that nuclear power is "the ultimate source of energy for mankind." (Cartledge 1993: 5)

"The fire at Chernobyl seemed to fulfil the green anxiety that nuclear might or might not be tolerable when it works, but when it goes wrong it is cataclysmic. Many people believe, contrary to all the evidence, that tens of thousands of people died as a result of the Chernobyl accident of April 1986". (North 1995: 82)

"We and our planet are wastes from a fusion reactor, the sun, . . . Scientists have found ways of restoring to its former vigour some of the dying radiation from the nuclear explosion which formed our world . . ." (ibid.: 84)

"The use of wind to generate electrical power is entering a period of massive growth, while the development of expensive nuclear energy is levelling off. The sale of fluorescent light bulbs which need far less electricity than incandescent ones has risen by 35% in the last five years. Lester R. Brown, president of the Worldwatch Institute, points out that his own office light bills have gone down by over 60% since switching to fluorescence." (*Herald*, 30 December 1993)

"For example, more that 90 per cent of people questioned were concerned about the destruction of the world's forests, and more than 80 per cent worried by threats of acid rain, global warming, and the destruction of the ozone layer. Air, sea, river and lake pollution all rated high on the list of public anxieties, and cars, taxis, buses and lorries were all regarded as big contributors to the problem. Nuclear power, and fear of radiation, prominent since the Chernobyl accident, remain a bogy in the public mind." (*Guardian*, 30 December 1993)

BOX A2.1 continued

"We already knew that 'out there' there are processes beyond the quiescent ones on Earth. The Sun is a nuclear fusion plasma ball; it is nuclear physics in action; it is replaying the processes that were going on everywhere throughout the universe when the universe was only minutes old. The extreme cosmic rays are hinting at conditions even more cataclysmic, such as those played out within the first seconds of the universe." (*Daily Telegraph*, 29 December 1993)

"The delusions of grandeur came under immediate assault. The landscape east of Paris is so flat, rain-smeared and dreary that you could, on the rebound, fall in love with Essex. It's hard to imagine anything you could build here that wouldn't improve it. An oil refinery, nuclear power station - anything. The notion that Euro Disney is some kind of visual excrescence, an American insult to pristine French scenery, is as wide of the mark as the idea that Frenchmen don't eat ketchup or can't identify with Donald Duck. 'A cultural Chernobyl' is how some members of the French government saw the place . . ." (*ibid.*, 23 December 1993)

"Many nuclear power plants have been thoughtfully sited in these earthquake zones already, simplifying the transport problems. Once at the bottom of a borehole, the waste will heat up the surrounding rock and melt its way downwards. Ultimately, it will reach the fault, where it will strongly heat up the local rock, melting or at any rate greatly weakening it." (*Guardian*, 23 December 1993)

"The regional, as opposed to local, environmental viability of electric cars depends almost exclusively on the types of energy used in the countries in question. In France, with only a ten per cent reliance on fossil-fuel energy, the electric car is a clear boon. In Germany, however, the balance becomes even more borderline as nuclear facilities are phased out." (*European*, 31 December 1993)

"Some Christmas trees in the former Soviet republic of Belarus have been found to be radioactive, apparently because of the 1986 disaster at the Chernobyl nuclear power plant, a new agency said Wednesday. The Belinform agency said many people preparing for the Russian Orthodox Christmas on Jan. 7 now prefer to buy artificial trees to avoid radiation dangers. A tree bought by one man from a roadside salesman contained 'unacceptably high levels of radiation,' it said. The man later tested the fir tree with a home dosimeter, and the device 'went off the wall'." (Associated Press, 29 December 1993)

Bibliography

Adam, B. 1994. Running out of time: global crisis and human engagement. In *Social theory and the global environment*, M. Redclift & T. Benton (eds), 92–112. London: Routledge.

Adger, N. 1993. Sustainable national income and natural resource degradation in Zimbabwe. See Turner (1993), 338–59.

Albert, B. 1993. Cannibal gold and the falling sky: a shamanic critique of the political economy of nature. *Homme* 33(2–4), 126–8, 349–78.

Alpert, P. 1993. Conserving biodiversity in Cameroon. *Ambio* 22(1), 44–9.

Alcamo, J. & B. De. Vries 1992. Low energy, low emissions: So_2, NO_x & CO_2 in western Europe. *International Environmental Affairs* 4(3), 155–84.

Andelson, R. V. 1991. Commons without tragedy: the congruence of Garret Hardin and Henry George. In *Commons without tragedy*, R. V. Andelson (ed.), 27–43. London: Shepheard–Welwyn.

Anderson, A. 1990. Smokestacks in the rainforest: industrial development and deforestation in the Amazon Basin. *World Development* 18(9), 1191–205.

Andrews, R. 1995. *Teaching and learning argument*. London: Cassell.

Apichatvullop, Y. & J. Compton 1993. Local participation in social forestry. *Regional Development Dialogue* 14(1), 34–42.

Aristotle 1926. *The art of rhetoric* [translated by J. H. Freese]. London: Heinemann.

Atkinson, M. 1984. *Our masters' voices*. London: Methuen.

Axelrod, R. 1992. Reconciling energy use with environmental protection in the European Community. *International Environmental Affairs* 4(3), 185–202.

Bakhtin, M. M. 1986. *Speech genres and other late essays* [V. McGee (trans.), C. Emerson & M. Holqvist (eds)]. Austin: University of Texas Press.

Balee, W. 1993. Indigenous transformation of Amazonian forests: an example from Maranhao, Brazil. *Homme* 33(2–4), 126–8, 231–54.

Barbosa, L. C. 1993. The world system and the destruction of the Brazilian Amazon rain forest. *Review* 16(2), 215–40.

Barker, T., S. Baylis, P. Madsen 1993. A UK carbon/energy tax. *Energy policy* 21(3), 296–308.

Barkham, J. 1995. Ecosystem management and environmental ethics. In *Environmental science for environmental management*, T. O'Riordan (ed.), 80–104. Harlow: Longman.

Barnett, R. 1995. *The limits of competence: knowledge, higher education and society*. Milton Keynes: Open University Press.

Baudrillard, J. 1994. *The illusion of the end*. Cambridge: Polity Press.

Bauman, Z. 1992. *Intimations of postmodernity*. London: Routledge.

Bazerman, C. 1987. Codifying the social scientific style: the APA *Publication manual* as a behaviourist rhetoric. In *The rhetoric of the human sciences*, J. Nelson et al. (eds), 125–44. Madison: University of Wisconsin Press.

Beckerman, W. 1994. "Sustainable development": is it a useful concept. *Environmental Values* 3, 191–209.

Bell, A. 1994. Climate of opinion – public and media discourse on the global environment. *Discourse and Society* 5(1), 33–64.

Berman, M. 1982. *All that is solid melts into air*. London: Verso.

Berry, W. 1990. *What are people for?* New York: North Point Press.

Bertram, G. 1992. Tradeable emission permits and the control of greenhouse gases. *Journal of Development Studies* 28(3), 423–46.

Beyea, J. 1990. Nuclear as last resort to change in climate. *Forum for Applied Research and Public Policy* 5(3), 90–92.

Biehl, J. 1991. *Finding our way: rethinking ecofeminist politics*. Montreal: Black Rose Books.

Billig, M. 1987. *Arguing and thinking: a rhetorical approach to social psychology*. Cambridge: Cambridge University Press.

—1991. *Ideology and opinions*. London: Sage.

—1992. *Talking of the royal family*. London: Routledge.

—1995. *Banal nationalism*. London: Sage.

Black, J., M. Levi, D. de Meza 1993. Creating a good atmosphere: minimum participation for tackling the "greenhouse effect". *Economica* 6(239), 281–93.

Blake, W. 1988(1790). The marriage of heaven and hell. In *William Blake*, M. Mason (ed.), 8–19. Oxford: Oxford University Press.

Blitzer, C. et al. 1993. Growth and welfare losses from carbon emission restrictions: a general equilibrium analysis for Egypt. *Energy Journal* 14(1), 57–81.

Bondi, H. 1993. Energy and the environment: the differing problems of the developed and developing regions. See Cartledge (1993), 11–23.

Bookchin, M. 1986. *The limits of the city*. Montreal: Black Rose Books.

—1991. *The ecology of freedom: the emergence and dissolution of hierarchy*. Montreal: Black Rose Books.

—1994. *Which way for the ecology movement?* Edinburgh: AK Press.

Bradshaw, Y. W. & E. E. Fraser 1990. City size, birth rates and development in China: evidence of modernisation. *Journal of Urban Affairs* 12(4), 401–424.

Bramwell, A. 1994. *The fading of the greens: the decline of environmental politics in the west*. New Haven: Yale University Press.

Brundtland, G. H. 1993. *Population, environment and development* [The Rafael M. Solas memorial lecture]. New York: United Nations.

Burgess, J. 1990. The production and consumption of environmental meanings in the mass media: a research agenda for the 1990s. *Transactions of the Institute of British Geographers* 15(2), 139–61.

Burke, K. 1950. *A rhetoric of motives*. Englewood Cliffs, New Jersey: Prentice-Hall.

—1989. *On symbols and society*. Chicago: University of Chicago Press.

—1989(1969). Vocabularies of motive. In *On symbols and society*, J. R. Gusfield (ed.), pp.158–76, Chicago: University of Chicago Press.

Callenbach, E. 1975. *Ecotopia*. Berkeley: Banyan Tree Books.

Campbell, J. A. 1987. Charles Darwin: rhetoricism of science. In *The rhetoric of the human sciences*, J. Nelson et al. (eds), 69–86. Madison: University of Wisconsin Press.

Capra, F. 1988. *Uncommon wisdom: conversations with remarkable people*. London: Century Hutchinson.

Carson, R. 1962. *Silent spring*. Harmondsworth: Penguin.

Cartledge, B. (ed.) 1992. *Monitoring the environment*. Oxford: Oxford University Press.

Cartledge, B. 1993. *Energy and the environment*. Oxford: Oxford University Press.

Casti, J. L. 1995. *Searching for certainty: what scientists can know about the future*. London: Abacus.

Catton, W. 1993. Carrying capacity and the death of a culture: a tale of two autopsies. *Sociological Inquiry* 63(2), 202–223.

Chester, P. 1993. Clean power from fossil fuels. See Cartledge (1993), 40–64.

Coldicutt, S., & T. Williams 1992. Concepts of solar energy use for climate control in buildings. *Energy Policy* 20(9), 825–35.

Common, M. S. 1989. The choice of pollution control instruments: why is so little notice taken of economists' recommendations? *Environment and Planning A* 21, 1297–314.

Commoner, B. 1991. The governance of production: the key to environmental quality. See Tolba & Biswas (1991: 95–102).

Connolly, W. E. 1984. The politics of discourse. In *Language and politics*, M. Shapiro (ed.), 139–67. Oxford: Basil Blackwell.

Connor, S. 1992. *Theory and cultural value*. Oxford: Basil Blackwell.

Conrad, J. 1974(1901). *Heart of darkness*. Harmondsworth: Penguin.

Corry, S. 1993. The rain forest harvest: who reaps the benefit? *The Ecologist* 23(4), 148–53.

Cummings, B. J. 1990. *Dam the rivers, damn the people: development and resistance in Amazonian Brazil*. London: Earthscan.

Daly, H. 1992. *Steady-state economics*, 2nd edn. London: Earthscan.

Darwin, C. 1959(1845). *The voyage of 'The Beagle'*. London: Dent.

—1968(1859). *The origin of species*. Harmondsworth: Penguin.

Dasgupta, C. 1994. The climate change negotiations. In *Negotiating climate change*, I. Mintzer & J. Leonard (eds), 129–48. Cambridge: Cambridge University Press.

Davis K. & M. Bernstam (eds) 1991. *Resources, environment and population: present knowledge, future options.* New York: Population Council and Oxford University Press.

Davis. P. & R. Hersch 1987. Rhetoric and matheratus. In *The rhetoric of the human sciences*, J. Nelson, A. Megill, O. N. McCloskey (eds), 53–68. Madison: University of Wisconsin Press.

Dawkins, R. 1982. *The extended phenotype.* Oxford: Oxford University Press.

Dear, M. 1995. Prologomena to a postmodern urbanism. In *Managing cities: the new urban context*, P. Healey, S. Cameron, S. Davoudi, S. Graham, A. Madani-Pour, 27–44. Chichester: John Wiley.

de la Court, T. 1990. *Beyond Brundtland: green development in the 1990s.* London: Zed Books/New York: New Horizons Press.

Del Amo, R. & P. Ramos 1993. Use and management of secondary vegetation in a humid-tropical area. *Agroforestry Systems* 21, 27–43.

DeLuchi, M. A. 1993. Greenhouse gas emissions from the use of new fuels for transportation and electricity. *Transportation Research A* 27(3), 187–219.

Department of the Environment/Welsh Office. 1993. *Integrated pollution control: a practical guide* London: HMSO.

Derrida, J. 1992. *Cinders* [translated by N. Lukacher]. Lincoln: University of Nebraska.

Devall, B. & G. Sessions 1985. *Deep ecology: living as if nature mattered.* Layton: Gibbs Smith.

Dixon, J., L. Fallon-Sur, T. Van't Hof 1993. Meeting ecological and economic goals: marine parks in the Caribbean. *Ambio* 22(2–3), 117–25.

Dobson, A. 1995. *Green political thought*, 2nd edn. London: Routledge.

Douglas, M. 1966. *Purity and danger.* London: Routledge & Kegan Paul.

Dowding, K. 1995. Model or metaphor? a critical review of the policy network approach. *Political Studies* 43(1), 136–58.

HRH Duke of Edinburgh 1991. Life on Earth. See Tolba & Biswas (1991), 18–24.

Duraiappah, A. 1993. *Global warming and economic development: a holistic approach to international policy cooperation and coordination.* Norwell, Mass.: Kluwer Academic.

Dyson, T. 1995. Loaves, fishes and cereal yields. *The Times Higher* (27 January), 17.

Easterley, C., T. Jones, L. Glass, B. Owens, P. Walsh 1993. Biotesting waste water for hazard evaluation. *Water Research* 27(7), 1145–52.

Eckersley, R. 1992. *Environmentalism and political theory: toward an ecocentric approach.* London: UCL Press.

The Ecologist 1993. *Whose common future? reclaiming the commons.* London: Earthscan.

Edwards, K. 1992. A new word order. In *Good science*. New York: Roof Books.

Egunjobi, L. 1993. Issues in environmental management for sustainable development in Nigeria. *The Environmentalist* 13(1), 33–40.

Ehrlich, P. & A. Ehrlich 1990. *The population explosion.* New York: Simon & Schuster.

Bibliography

Ehrlich, P., A. Ehrlich, G. Daily 1993. Food security, population and environment. *Population and Development Review* 19(1), 1–32.

Ehrlich, P. & E. Wilson 1991. Biodiversity studies: science and policy. *Science* 253, 758–62.

Ekins, P. 1992. *A new world order: grassroots movements for global change.* London: Routledge.

Ekins, P. & M. Max-Neef (eds) 1992. *Real life economics.* London: Routledge.

Ekins, P., M. Hillman, R. Hutchinson 1992. *The Gaia atlas of green economics.* New York: Anchor. Books.

Elster, J. 1989. *Solomonic judgements.* Cambridge: Cambridge University Press.

Elton, B. 1989. *Stark.* London: Warner.

Farrenkopf, J. 1993. Spengler's historical pessimism and the tragedy of our age. *Theory and Society* 22(3), 391–412.

Featherstone, M. (ed.) 1992. *Cultural theory and cultural change.* London: Sage.

Feeney, G. 1991. Comment: on the uncertainty of population projections. In *Resources, environment and population: present knowledge, future options,* K. Davis & M. Bernstam (eds), 72–5. New York: Population Council and Oxford University Press.

Fischer, F. & J. Forester (eds) 1993. *The argumentative turn in policy analysis and planning.* London: UCL Press.

Fish, S. 1989. *Doing what comes naturally.* Oxford: Oxford University Press.

Flavin, C. & N. Lenssen 1992. Policies for a solar economy. *Energy Policy* 20(3), 245–56.

Foster, J. B. 1993. "Let them eat pollution": capitalism and the world environment. *Monthly Review* 44(8), 10–20.

Fukuyama, F. 1992. *The end of history and the last man.* London: Penguin.

Gable, E. J., D. G. Aubrey, J. H. Gentile 1991. Global environmental change issues in the western Indian Ocean region. *Geoforum* 22(4), 401–419.

Gadamer, H-G. 1979. *Truth and method.* London: Sheed & Ward.

Gaillie W. B. 1968. *Philosophy and the historical understanding.* New York: Schocken Books.

Gardner, R. 1992. *Negotiating survival: four priorities after Rio.* New York: Council on Foreign Relations Press.

Gayoom, M. A. 1991. Climatic change, environment and development. See Tolba & Biswas (1991: 14–17).

Geertz, C. 1973. *The interpretation of cultures.* New York: Basic Books.

Giddens, A. 1990. *The consequences of modernity.* Cambridge: Polity.

—1991. *Modernity and self-identity: self and society in the late modern age.* Cambridge: Polity.

—1992. *The transformation of intimacy.* Cambridge: Polity.

—1994. *Beyond left and right: the future of radical politics.* Cambridge: Polity.

Giddings, L. V. & M. Garcia-Castro 1991. Biotechnology and biodiversity. *Sociologica* 6(16), 273–303.

Gilligan, C. 1982. *In a different voice: psychological theory and women's devel-*

opment. Cambridge, Mass.: Harvard University Press.

Girardet, H. 1992. *Earthrise: how we can heal our injured plant*. London: Paladin.

Goodison, L. 1994. Were the Greeks green. In *Green history*, D. Wall (ed.), 179–81. London: Routledge.

Goodrich, P. 1986. *Reading the law: a critical introduction to legal method and techniques*. Oxford: Basil Blackwell.

Gordon, D. M. 1993. ECO-ED: rhetoric or progress. *Futures* 25(1), 98–100.

Gore, A. 1993. *Earth in the balance: ecology and the human spirit*. New York: Penguin Books.

Gould, S. J. 1989. *Wonderful life: the Burgess shale and the nature of history*. Harmondsworth: Penguin.

Grant, A. 1995. Human impacts on terrestrial ecosystems. In *Environmental science for environmental management*, T. O'Riordan (ed.), 66–79. London: Longman.

Gray, A. 1991. The impact of biodiversity conservation on indigenous peoples. In *Biodiversity: social and ecological perspectives*, V. Shiva et al. (eds), 59–76. London: Zed Books with World Rainforest Movement.

Gross, A. 1989. *The rhetoric of science*. Harvard: CUP.

Grove-White, R. 1993. Environmentalism: a new moral discourse for technological society? In *Environmentalism: the view from anthropology*, K. Milton (ed.), 18–30. London: Routledge.

Grove-White, R. & B. Szerszynski 1992. Getting behind environmental ethics. *Environmental Values* 1(4), 285–96.

Grubb, M., M. Koch, A. Manson, F. Sullivan, K. Thomson 1993. *The Earth summit agreements: a guide and assessment*. London: Royal Institute of International Affairs / Earthscan.

Habermas, J. 1984. *The theory of communicative action*, vol.1: *reason and the rationalisation of society* [translated by T. McCarthy]. Cambridge: Polity.

—1986. *Philosophical discourse of modernity*. Cambridge: Polity.

—1987. *The theory of communicative action*, vol. 2: *lifeworld and system: the critique of functionalist reason* [translated by T. McCarthy]. Cambridge: Polity.

Hammer, M., A. Jansson, B. Jannson 1993. Diversity change and sustainability: implications for fisheries. *Ambio* 22(2–3), 97–105.

Haraway, D. 1991. *Simians, cyborgs and women: the reinvention of nature*. London: Free Association Books.

Hardin, G. 1968. The tragedy of the commons. *Science* 162, 1243–8. [Reprinted in *An annotated reader in environmental planning and management*, T. O'Riordan & R. K. Turner (eds), 288–99. Oxford: Pergamon, 1983.]

—1991. The tragedy of the *unmanaged* commons: population and the disguises of providence. See Andelson (1991: 162–85).

Hargrove, E. (ed.) 1992. *The animal rights / environmental ethics debate: the environmental perspective*. Albany: State University of New York Press.

Harrington, J., D. Wilcox, P. Giles, N. Ashbolt, J. Evans, H. Kirton 1993. The health of Sydney surfers: an epidemiological study. *Water, Science and*

Technology 27(3–4), 175–81.

Harrison C. & J. Burgess 1994. Social construction of nature: a case study of conflicts over the development of Rainham Marshes. *Transactions of the Institute of British Geographers* 19(3), 291–310.

Harrison, E. 1991. The crisis of transition from the commons: population explosions, their cause and issue. See Andelson (1991: 44–82).

Hart, D. & D. Victor 1993. Scientific elites and the making of US policy for climate change research, 1957–74. *Social Studies of Science* 23(4), 643–80.

Hart, T. 1994. Transport choices and sustainability: a review of changing trends and policies. *Urban Studies* 31(4/5), 705–27.

Heidegger, M. 1993. *Basic writings* [edited by D. Farrell Krell]. London: Routledge.

Henderson, H. 1993. *Paradigms in progress: life beyond economics*. London: Adamantine Press.

Hilton, M. J. 1994. Applying the principle of sustainability to coastal sand mining. *Environmental Management* 18(6), 814–29.

Hird, J. 1993. Environmental policy and equity: the case of superfund. *Journal of Policy Analysis and Management* 12(2), 323–43.

Hirschman, A. O. 1991. *The rhetoric of reaction*. Cambridge, Mass.: Belknap Press.

Hobsbawm, E. 1968. *Industry and empire*. Harmondsworth: Penguin.

—1994. *Age of extremes: the short twentieth century, 1914–91*. London: Michael Joseph.

Horn, R. 1993. *Statistical indicators for the economic and social sciences*. Cambridge: Cambridge University Press.

Hughes, T. 1994. Creation/Four Ages/Flood. In *After Ovid: new metamorphoses*, M. Hofman & J. Lasdun (eds), 3–20. London: Faber & Faber.

Hyder, T. O. 1992. Client negotiations: the north/south perspective. See Mintzer (1992), 323–36.

Ince, M. 1990. *The rising seas*. London: Earthscan.

Independent Commission on International Humanitarian Issues 1986. *The vanishing forest*. London: Zed Books.

Ingold, T. 1993. Globes and spheres: the topology of environmentalism. In *Environmentalism: the view from anthropology*, K. Milton (ed.), 31–42. London: Rout-ledge.

Inman, K. 1993. Fuelling expansion in the Third World: population, development, debt and the global decline of forests. *Society and Natural Resources* 6(1), 17–39.

ITE/NERC 1989. *Climatic change, rising sea level and the British coast*. London: HMSO.

IPCC 1990. *Policymakers' summary on the scientific assessment of climate change*. WMO/UNEP.

Jacobs, M. 1991. *The green economy*. London: Pluto.

—1994. The limits to neo-classicalism: towards an institutional environmental economics. In *Social theory and the global environment*, M. Redclift & T. Benton (eds), 67–91. London: Routledge.

Jarallah, J., K. Noweir, S. Al-Shammari, I. Al-Saleh, M. Al-Zahrani, M. Al-Aged 1993. Lead exposure among school children in Ryiadh, Saudi Arabia. *Bulletin of Environmental Contamination Toxicology* 50(5), 730–35.

Jardine, K. 1994. Finger on the carbon pulse: climate change and the boreal forests. *The Ecologist* 24(6), 220–23.

Johnson, L. E. 1991. *A morally deep world: an essay on moral significance and environmental ethics*. Cambridge: Cambridge University Press.

Johnson-Freese, J., R. Handberg, D. C. Webb 1992. Return from orbit: economics as a driver of Japanese space policy. *Technology in Society* 14, 395–408.

Kempton, W. 1991. Lay perspectives on global climate change. *Global Environmental Change* 1(3), 183–208.

Kennedy, G. A. 1963. *The art of persuasion in ancient Greece*. London: Routledge & Kegan Paul.

Kennedy, G. A. 1994. *A new history of classical rhetoric*. Princeton, New Jersey: Princeton University Press.

Killingsworth, M. & J. Palmer 1992. *Ecospeak*. Carbondale: Southern Illinois University Press.

Klaiss, H. & C. J. Winter 1992. Systems comparison and potential of solar thermal installations in the Mediterranean area. *Revue de l'Energie* 441, 602–18.

Klamer, A. 1988. Negotiating a new conversation about economics. In Klamer et al. (1988: 265–79).

Klamer, A. & D. N. McCloskey 1988. Economics in the human conversation. In Klamer et al. (1988: 3–20).

Klamer, A., R. M. Solow, D. N. McCloskey 1988. *The consequences of economic rhetoric*. New York: Cambridge University Press.

Kuhn, T. S. 1970. *The structure of scientific revolutions*. Chicago: University of Chicago Press.

Lacey, C. & D. Longman 1993. The press and public access to the environment and development debate. *The Sociological Review* 41(2), 207–43.

Lakoff, G. & M. Johnson 1980. *Metaphors we live by*. Chicago: University of Chicago Press.

Lakoff, G. & M. Turner 1989. *More than cool reason: a field guide to poetic metaphor*. Chicago: University of Chicago Press.

Lalonde, B. 1991. Protecting the atmosphere and controlling global warming. See Tolba & Biswas (1991: 82–7).

Lanham, R. A. 1968. *A handlist of rhetorical terms*. Berkeley: University of California Press.

Lave, L. & K. Vickland 1989. Adjusting to greenhouse effects: the demise of traditional cultures and the cost to the USA. *Risk Analysis* 9(3), 283–91.

Lee, R. 1991. Long run global population forecasts: a critical appraisal. In *Resources, environment and population: present knowledge, future options*, K. Davis & M. Bernstam (eds), 44–71. New York: Population Council and

Oxford University Press.

Leff, M. 1987. Modern sophistic and the unity of rhetoric. In *The rhetoric of the human sciences*, J. Nelson, A. Megill, D. McCloskey (eds), 19–37. Madison: University of Wisconsin Press.

Le-Houerou, H. N. 1990. Global change: population land-use and vegetation in the Mediterranean basin by the mid-21st century. In Paepe et al. (1990: 301–367).

Leggett, J. 1992. Running down to Rio. *New Scientist* 134(2 May), 38–42.

—1993. Anxieties and opportunities in climate change. See Prins (1993), 41–65.

Leith, D. & G. Myerson 1989. *The power of address: explorations in rhetoric.* London: Routledge.

Lenssen, N. 1993. All the coal in China. *World Watch* 6(2), 22–9.

Leopold, A. 1949. *A sand county almanac.* New York: Oxford University Press. See Wall (1994: 75–6).

Lessing, D. 1979. *Shikasta.* London: Grafton.

Lewis, M. 1994. *Green delusions: an environmentalist of radical environmentalism.* Durham: Duke University Press.

Lindblom, C. E. 1990 *Inquiry and change: the troubled attempt to understand and shape society.* New Haven, Connecticut: Yale University Press.

Livingstone, S. & P. Lunt 1993. *Talk on television.* London: Routledge.

Lloyd, G. 1993. *Being in time: selves and narrators in philosophy and literature.* London: Routledge.

Lone, Ø. 1992. Environmental and resource accounting. In *Real-life economics: understanding wealth creation*, P. Ekins & M. Max-Neef (eds), 239–54. London: Routledge.

Lovelock, J. 1979. *Gaia: a new look at life on Earth.* Oxford: Oxford University Press.

—1988. *The ages of Gaia: a biography of our living Earth.* Oxford: Oxford University Press.

—1992. The Earth is not fragile. In *Monitoring the environment*, B. Cartledge (ed.), 105–22. Oxford: Oxford University Press.

Lozada, G. 1993. The conservationists's dilemma. *International Economic Review* 34(3), 647–62.

Lugo, A., J. Parrotta, S. Brown 1993. Loss in species caused by tropical deforestation and their recovery through management. *Ambio* 22(2–3), 106–9.

Luteyn, J. L. 1992. Paramos: why study them? In *Parano: an Andean ecosystem under human influence*, H. Balsler & J. Luteyn (eds), 1–14. London: Academic Press.

Lutz, W. & J. Baguart 1992. Population and sustainable development: a case study of Mauritius. See van Imhoff et al. (1992: 57–82).

MacGillivray, A. & S. Zadek 1995. Accounting for change. London: New Economics Foundation.

McCloskey, D. N. 1985. *The rhetoric of economics.* Madison: University of Wisconsin Press.

—1990. *If you're so smart: the narrative of economic expertise.* Chicago: Chicago

The language of environment

University Press.

—1994. *Knowledge and persuasion in economics*. Cambridge: Cambridge University Press.

McCormick, F. 1992. Night comes to Amazonia. *Forum for Applied Research and Public Policy* 7(4), 30–4.

McKendry, J. & G. Machlis 1991. The role of geography in extending biodiversity gap analysis. *Applied Geography* 11, 135–52.

McKibben, B. 1992. An interview with David Brouwer. In *The rolling stone environmental reader*, 235–48. Washington DC: Island Press.

McLuhan, M. 1967. *The medium is the message*. (Harmondsworth: Penguin)

McNamara, R. S. 1991. The population problem. See Tolba & Biswas (1991: 48–65).

McNealey, J. 1993. Economic incentives for conserving biodiversity: lessons for Africa. *Ambio* 22(2–3), 144–50.

Malthus, T. 1970(1798). *An essay on the principle of population*. Harmondsworth: Penguin.

Malueg, D. A. 1990. Welfare consequences of emission credit trading programs. *Journal of Environmental Economics and Management* 18(1), 66–77.

Maunder, P., D. Myers, N. Wall, R. Le Roy Miller 1991. *Economics explained*, 2nd edn. London: Collins Educational.

Manne, A. & R. Richels 1993. The EC proposal for combining carbon and energy taxes. *Energy Policy* 21(1).

Margulis, L. & G. Hinkle 1991. The Biota and Gaia: 150 years of support for environmental sciences. In *Scientists on Gaia*, S. Schneider, & P. Boston (eds), 11–18. Cambridge, Mass.: MIT Press.

Markandya, A. & J. Richardson (eds) 1992. *The Earthscan reader in environmental economics*. London: Earthscan.

Marshall, W. 1993. Nuclear power and the environment. See Cartledge (1993: 146–58).

Martin, E. B. 1992. The poisoning of rhinos and tigers in Nepal. *Oryx* 26(2), 82–6.

Martin, P. & S. Lockie 1993. Environmental information for total catchment management. *Australian Geographer* 24(1), 75–85.

Martin, W. E., R. H. Patrick, B. Tolwinski 1993. A dynamic game of a transboundary pollutant with asymmetric players. *Journal of Environmental Economics and Management* 25(1), 1–12.

Martinez-Alier, J. 1987. *Ecological economics: energy environment and society*. Oxford: Basil Blackwell.

Massiah, J. (ed.) 1993. *Women in developing economies: making visible the invisible*. Oxford: Berg/UNESCO.

Matthews, R. 1994. The rise and rise of global warming. *New Scientist* (26 November), 6.

Mazur, A. & J. Lee, 1993. Sounding the global alarm: environmental issues in the us national news. *Social Studies of Science* 23, 681–720.

Megill, A. & D. N. McCloskey 1987. The rhetoric of history. In *The rhetoric of the human sciences*, J. Nelson et al. (eds), 221–38. Madison: University of Wisconsin Press.

246

Meier-Braun, K. H. 1992. The new mass migration. *Scala* **9–10**, 10–13.

Merchant, C. 1980. *The death of nature*. San Francisco: Harper and Row.

Midgley, M. 1992. The significance of species. In *The animal rights / environmental ethics debate*, E. Hargrove (ed.), 121–36. Albany: State University of New York Press.

Mill, J. S. 1974. *On liberty*. Harmondsworth: Penguin.

Miller, C. & S. Halloran 1993. Reading Darwin, reading nature; or, on the ethos of historical science. In *Understanding scientific prose*, J. Selzer (ed.), 106–26. Madison: University of Wisconsin Press.

Miller, P. 1994. Accounting and objectivity: the invention of calculating selves and calculable spaces. In Rethinking objectivity, A. Megill (ed.), 239–64. Durham, North Carolina: Duke University Press.

Miller, S. 1991. *Textual carnivals: the politics of composition*. Carbondale: Southern Illinois University Press.

Milton, K. (ed.) 1993. *Environmentalism: the view from anthropology*. London: Routledge.

Mintzer, I. M. (ed.) 1992. *Confronting climate change*. Cambridge: Cambridge University Press.

Mintzer, I. M. & J. A. Leonard 1994. *Negotiating climate change: the inside story of the Rio Convention*. Cambridge: Cambridge University Press.

Mitchell, B. & J. Barborak 1991. Developing coastal park systems in the tropics: planning in the Turks and Caicos Islands. *Coastal Management* **19**(1), 113–34.

Mott, J. & P. Bridgewater 1992. Biodiversity conservation and ecologically sustainable development. *Search* **23**(9), 284–7.

Mugabe, R. 1991. Environmental concerns in the Third World. See Tolba and Biswas (1991), 10–3.

Mukherjee, N. 1992. Greenhouse gas emissions and the allocation of responsibility. *Environment and Urbanization* **1**(1), 89–98.

Murdoch, J. & J. Clark 1994. Sustainable knowledge. *Geoforum* **25**(2), 115–32.

Murray, M. 1993. The value of biodiversity. See Prins (1993), 66–86.

Myers, N. 1992. Population/environment linkages: discontinuities ahead? See van Imhoff et al. (1992), 15–31.

—1993. *Ultimate security: the environmental basis of political stability*. New York: W. W. Norton.

—1994. Editorial. *The Environmentalist* **14**(2), 86.

Myers, N. & J. Simon 1994. *Scarcity or abundance? a debate on the environment*. New York: W. W. Norton.

Myerson, G. 1992. *The argumentative imagination*. Manchester: Manchester University Press.

—1994. *Rhetoric, reason and society*. London: Sage.

Myerson, G. & Y. Rydin 1994. "Environment" and planning: a tale of the mundane and the sublime. *Environment and planning D* **12**, 437–52.

Myerson, G. & Y. Rydin 1995. Sustainable development: the implications of the global debate for land use planning. In *Environmental planning and sustainability*, B. Evans & S. Buckingham-Hatfield (eds), 19–34. Chichester: John Wiley.

Naess, A. 1994. Sustainable development and the deep ecology movement. Paper to European Consortium for Political Research conference on The Politics of Sustainable Development within the European Union. University of Crete.

NAS (National Academy of Sciences) 1990. *One earth, one future: our changing global environment*. Washington: National Academy Press.

Nordhaus, W. D. 1993. Rolling the "dice": an optimal transition path for controlling greenhouse gases. *Resource and Energy Economics* 15(1), 27–50.

Norgaard, R. 1992. Coevolution of economy, society and environment. In *Real-life economics: understanding wealth creation*, P. Ekins and M. Max-Neef (eds), 76–88. London: Routledge.

North, R. 1995. *Life on a modern planet: a manifesto for progress*. Manchester: Manchester University Press.

Nussbaum, M. 1990. *Love's knowledge: essays on philosophy and literature*. New York: Oxford University Press.

Oates, W., P. Portney, A. McGartland. n.d. *The NET benefits of incentive-based regulation: the case of environmental standard setting in the real world*. Working Paper Series 88–33. Department of Economics, University of Maryland.

OECD 1975. *The polluter pays principle: definition, analysis, implementation*. Paris: OECD.

O'Riordan, T. 1988. The politics of sustainability. In *Sustainable environmental economics and management: principles and practice*, R. K. Turner (ed.), 29–50. London: Pinter (Belhaven).

—1992. The environment. In *Policy and change in Thatcher's Britain*, P. Cloke (ed.), 297–324. Oxford: Pergamon.

—1993. The politics of sustainability. In *Sustainable environmental economics and management: principles and practice* (2nd edn), R. K. Turner (ed.), 37–69. London: Pinter (Belhaven).

—(ed.) 1995. *Environmental science for environmental management*. London: Longman.

O'Rourke, P. J. 1992. The greenhouse affect. In *The rolling stone environmental reader*, foreword by James B. Wenner, 15–24. Washington DC: Island Press.

Overseas Development Agency 1991. *Population, environment and development*. An issues paper for the 3rd UNCED Prep Comm. London: ODA.

Paepe, R. et al. 1990. *Greenhouse effect, sea level and drought*. Proceedings of workshop, Fuerteventura, 1989. Kluwer for NATO ASI series; series C, 325.

Paoletti, M., D. Pimentel, B. Stinner, D. Stinner 1992. Agroecosystem, biodiversity: matching production and conservation biology. *Agriculture, Ecosystems and Environment* 40, 3–23.

Pearce, D. 1993. *Blueprint 3: measuring sustainable development*. London: Earthscan.

—1995. *Blueprint 4: capturing global environmental value*. London: Earthscan.

Pearce, D., A. Markandya, E. Barbier 1989. *Blueprint for a green economy*. London: Earthscan.

Pearce, D. & R. K. Turner 1990. *Economics of natural resources and the envi-*

ronment. Hemel Hempstead: Harvester Wheatsheaf.

Pearce, D. & D. Moran, with IUCN (The World Conservation Union) 1994. *The economic value of biodiversity.* London: Earthscan.

Pearce, F. 1994a. Will global warming plunge Europe into an ice age. *New Scientist* 144(19 November), 1952, 20–1.

——1994b. Political paralysis stalls biodiversity talks. *New Scientist* 144, 5.

Perelman, C. 1982. *The realm of rhetoric.* Notre Dame: University of Notre Dame Press.

Perelman, C. & L. Olbrechts-Tyteca 1969. *The new rhetoric: a treatise on argumentation.* Notre Dame, Indiana: University of Notre Dame Press.

Pestel, E. 1989. *The limits to growth.* New York: Universe Books.

Philipp, R. & A. Bates 1992. Health risks assessment of dinghy sailing in Avon and exposure to cynnobacteria (blue-green algae). *Journal of the Institution of Water and Environmental Management* 6(5), 613–20.

Phillips, V., A. Chuveliov, P. Takahashi 1992. A case study of renewable energy for Hawaii. *Energy* 17(2), 191–200.

Pinchot, G. 1901. *The fight for conversation.* Garden City, New York: Harcourt Brace. In *Green history*, D. Wall (ed.), 136. London: Routledge.

Porritt, J. 1993. Sustainable development: panacea, platitude or downright deception. See Cartledge (1993), 24–39.

Prance, G. 1990. Future of the Amazonian rainforest. *Futures* (November), 891–903.

HRH Prince of Wales 1993. "First unshackle the spirit". The Brundtland speech, April 1992. See Prins (1993), 3–14.

Prins, G. (ed.) 1993. *Threats without enemies: facing environmental insecurity.* London: Earthscan.

Putnam, H. 1981. *Reason, truth and history.* Cambridge: Cambridge University Press.

——1987. *The many faces of realism.* La Salle, Illinois: Open Court.

——1990. *Realism with a human face.* Cambridge, Massachusetts: Harvard University Press.

Qu-Geping 1992. China's dual-thrust energy strategy: economic development and environment protection. *Energy Policy* 20(6), 500–6.

Rae, J. 1993. Alternative energy sources: the European story. See Cartledge (1993), 92–106.

Rayner, S. 1993. Prospects for CO_2 emissions reduction policy in the USA. *Global Environmental Change* 3(1), 12–31.

RCEP 1988. *Best practicable environmental option*, 12th report [Cm 310]. London: HMSO.

——1994. *Transport and the environment*, 18th report [Cm 2674]. London: HMSO.

Read, P. 1994. *Responding to global warming.* London: Zed Books.

Redclift, M. 1994. Sustainable development and local empowerment: sources and sink. Paper to ECPR conference on The Politics of Sustainable Development within the European Union. University of Crete.

Redclift, M. & T. Benton (eds) 1994. *Social theory and the global environment*. London: Routledge.

Repetto, R., W. Magrath, M. Wells, L. Beer, F. Rossini 1992. Wasting assets: natural resources in the national income accounts. See Markandya & Richardson (1992: 364–88).

Rees, J. 1985. *Natural resources: allocation, economics and policy*. London: Methuen.

Renzoni, A. 1992. Comparative observations on levels of mercury in scalp hair of humans from different islands. *Environmental Management* 16(5), 597–602.

Robinson, K. S. 1994. *Green Mars*. London: HarperCollins.

Rohlman, M. 1993. Integrated natural resource management – a question of property institutions? *Geojournal* 29(4), 405–12.

Rooke, D. 1993. The gas industry: survivor and innovator. See Cartledge (1993), 15–91.

Rorty, R. 1987. Science as solidarity. In *The rhetoric of the human sciences*, J. Nelson et al. (eds), 38–52. Madison: University of Wisconsin Press.

Rose, C. 1990. *The dirty man of Europe*. London: Simon & Schuster.

Rosen, R. 1994. Who gets polluted? *Dissent* (Spring), 223–30.

Rosenhead, J. (ed.) 1989. *Rational analysis for a problematic world*. Chichester: John Wiley.

Roszak, T. 1972. *Where the wasteland ends: politics and transcendence in post-industrial society*. New York: Anchor Books.

Rothman, D. & D. Chapman 1993. A critical analysis of climate change policy research. *Contemporary Policy Issues* 11, 88–98.

Rydin, Y. 1993. *The British planning system*. London: Macmillan.

Sawyer, S. W. 1982. Leaders in change: solar energy owners and the implications for future adoption rates. *Technological Forecasting and Social Change* 21(3), 201–11.

Schneider, S. & P. Boston 1991. *Scientists on Gaia*. Cambridge, Mass.: MIT Press.

Schrag, C. 1992. *The resources of rationality: a response to the postmodern challenge*. Bloomington: Indiana University Press.

Schücking, H. & P. Anderson 1991. Voices unheard and unheeded. In *Biodiversity*, V. Shiva et al., 13–42. London: WRM & Zed Books.

Selzer, J. (ed.) 1993. *Understanding scientific prose*. Madison: University of Wisconsin Press.

Sen, A. 1985. Rationality and uncertainty. *Theory and Decision* 18, 109–27.

—1987. *On ethics and economics*. Oxford: Basil Blackwell.

Senior, O. 1994. *Gardening in the tropics*. Toronto: McClelland & Stewart.

Shaomin-Li 1989. China's population control and socio-economic reforms: the Chinese dilemma. *China Report* 25(3), 219–35.

Shiva, V. 1988. *Staying alive: women, ecology and development*. London: Zed Books.

Shiva, V., P. Anderson, H. Schücking, A. Gray, L. Lohmann, D. Cooper 1991. *Biodiversity: social and ecological perspectives*. London: Zed Books (with World

Rainforest Movement, Penang).

Shiva, V. & R. Holla-Bhar 1993. Intellectual piracy and the neem tree. *The Ecologist* 23(6), 223–7.

Shrivastava, P. 1993. Crisis theory/practice: towards a sustainable future. *Industrial and Environmental Crisis Quarterly* 7(1), 23–42.

Singer, P. 1990. *Animal liberation*. New York: Avon Books.

Smyth, I. 1995. Population policies: official responses to feminist critiques. Discussion Paper 14. Centre for the Study of Global Governance, London School of Economics.

Southgate, D. & M. Whitaker 1994. *Economic progress and the environment*. Oxford: Oxford University Press.

Southwood, R. 1992. The environment: problems and prospects. In *Monitoring the environment*, B. Cartledge (ed.), 5–41.

Stavins, R. & B. Whitehead 1992. Dealing with pollution. *Environment* 34(7), 6–11, 29–42.

Stocking, M. 1995. Soil erosion and land degradation. In *Environmental science for environmental management*, T O'Riordan (ed.), 223–42. London: Longman.

Swales, J. M. 1990. *Genre analysis: English in academic and research settings*. Cambridge: CUP.

Tahmassebi, C. 1992. Government role in achieving environmental goals: market forces versus regulations. *Energy Policy* 20(10), 959–62.

Tahvonen, O. & J. Kuuluvainen 1991. Economic growth, pollution and renewable resources. *Journal of Environmental Economics and Management* 24(2), 101–18.

Tainer, E. M. 1993. *Using economic indicators to improve investment analysis*. New York: John Wiley.

Tan Kim Yang, V. & J. M. Fox 1993. Participatory land use planning as a sociological methodology for natural resource management. *Regional Development Dialogue* 14(1), 70–83.

Taylor, P. & F. Buttel 1992. How do we know we have global environmental problems? science and the globalisation of environmental discourse. *Geoforum* 23(3), 405–16.

Thomas, V. & P. Kevan 1993. Basic principles of agroecology and sustainable agriculture. *Journal of Agricultural and Environmental Ethics* 6(1), 1–19.

Thompson, I. & P. Welsh 1993. Integrated resource management in boreal forest ecosystems – impediments and solutions. *The Forestry Chronicle* 69(1), 32–9.

Throgmorton, J. A. 1993. Survey research as rhetorical trope: electric power planning arguments in Chicago. In *The argumentative turn in policy analysis and planning*, F. Fischer & J. Forester (eds), 117–44. London: UCL Press.

Tickell, C. 1992. Implications of global climate change. In *Monitoring the environment*, B. Cartledge (ed.), 93–104. Oxford: OUP.

—1993. The inevitability of environmental insecurity. See Prins (1993), 15–26.

Tidmarsh, K. 1993. Russia's work ethic. *Foreign Affairs* 72(2), 67–77.

Tisdell, C. 1993. Project appraisal, the environment and sustainability for small island. *World Development* 21(2), 213–19.

Tolba, M. K. & A. K. Biswas (eds) 1991. *Earth and us: population – resources – environment –development*. London: Butterworth–Heinemann.

Tuchman Matthews, J. 1993. Nations and nature: a new view of security. See Prins (1993), 25–40.

Turner, R. K. 1993a. Sustainability: principles and practice. See Turner (1993), 3–36.

—(ed.) 1993b. *Sustainable environmental economics and management: principles and practice*. London: Belhaven.

Twitchell, K. 1991. The not-so-pristine Arctic. *Canadian Geographic* 111(1), 53–60.

UK Government 1994. *Sustainable development: the UK strategy*. London: HMSO.

van Breemen, N. 1993. Soils as biotic constructs favouring net primacy productivity. *Geoderma* 53, 183–211.

Vandermeer, J. 1991. The political ecology of sustainable development: the southern Atlantic coast of Nicaragua. *Centennial Review* 35(2), 265–94.

van Imhoff, E., E. Themmen, F. Willekens (eds) 1992. *Population, environment and development*. Amsterdam: Swets & Zeitlinger.

Vavrousek, J. 1993. Institutions for environmental security. See Prins (1993), 87–112.

Vellvé, R. 1993. The decline of diversity in European agriculture. *The Ecologist* 23(2), 64–9.

Vickers, B. 1988. *In defence of rhetoric*. Oxford: Oxford University Press.

Vincentnathan, L. 1993. Untouchable concepts of person and society. *Contributions to Indian Sociology* 27(1), 53–82.

Wall, D. 1994. *Green history: a reader in environmental literature, philosophy and politics*. London: Routledge.

Walton, A. L. & D. C. Hall 1990. Solar power. *Contemporary Policy Issues* 8, 240–54.

Walton, D. N. 1989 *Informal logic: a handbook for critical argumentation*. Cambridge: Cambridge University Press.

Ward, B. 1976. *The home of man*. Harmondsworth: Penguin.

Ward, B. & R. Dubos 1972. *Only one earth: the care and maintenance of a small planet*. Harmondsworth: Penguin.

Ward, H. 1993. Game theory and the politics of the global commons. *Journal of Conflict Resolution* 37(2), 203–35.

Watkins, M., T. Mason, P. Goodwin 1993. The role of geology in the development of Maputaland, South Africa. *Journal of African Earth Sciences* 16(1–2), 205–21.

Wells, M. & K. Brandon 1993. The principles and practice of buffer zones and local participation in biodiversity conservation. *Ambio* 22(2–3), 157–62.

Whatmore, S. 1995. *Possessing the earth: property and environment in late modernity*. London: Routledge.

Bibliography

Whatmore, S. & S. Boucher 1993. Bargaining with nature – the discourse and practice of environmental planning gain. *Transactions of the Institute of British Geographers* 18(2), 166–78.

White, H. 1978. *Tropics of discourse: essays in cultural criticism*. Baltimore: Johns Hopkins University Press.

Williams, R. 1981. *Culture*. London: Fontana.

Willis, K. & J. Benson 1993. Valuing environmental assets in developed countries. See Turner (1993), 269–95.

Worcester, R. 1993. *Societal values and attitudes to human dimensions of global environmental change*. Paper to International Conference on Social Values, Complutense University of Madrid, 28 September.

World Bank 1993. *Social indicators of development*. Baltimore: John Hopkins University Press.

World Commission on Environment and Development 1987. *Our common future*. Oxford: Oxford University Press.

Wortham, R. A. 1993. Population growth and demographic transition in Kenya. *International Sociology* 8(2), 197–214.

Wynne, B. 1994. Scientific knowledge and the global environment. In *Social theory and the global environment*, M. Redclift & T. Benton (eds), 169–89. London: Routledge.

Wysham, D. 1994. 10–1 against: costing people's lives for climate change. *The Ecologist* 24(6), 204–206.

Yamaguchi, K. 1993. Information-decision structures and futures research. *Futures* (Jan/Feb), 66–80.

Yearley, S. 1993. Standing in for Nature: the practicalities of environmental organizations' use of science. In *Environmentalism: the view from anthropology*, K. Milton (ed.), 59–72. London: Routledge.

Young, J. 1990. *Post-environmentalism*. London: Belhaven.

Zimmerman, M. E. 1990. *Heidegger's confrontation with modernity: technology, politics, art*. Bloomington: Indiana University Press.

Index

Page numbers printed in **bold type** refer to boxed text

Index

Index

Index